Microplastic sources, fate and solution

Anish Khan • Chongqing Wang •
Abdullah M. Asiri
Editors

Microplastic sources, fate and solution

Springer

Editors
Anish Khan
Center of Excellence for Advanced
Materials Research
King Abdulaziz University
Jeddah, Saudi Arabia

Chongqing Wang
School of Chemical Engineering
Zhengzhou University
Zhengzhou City, China

Abdullah M. Asiri
Center of Excellence for Advanced
Materials Research
King Abdulaziz University
Jeddah, Saudi Arabia

Chemistry Department, Faculty of Science
King Abdulaziz University
Jeddah, Saudi Arabia

ISBN 978-981-99-0694-9 ISBN 978-981-99-0695-6 (eBook)
https://doi.org/10.1007/978-981-99-0695-6

This Springer imprint is published by the registered company Springer Nature Singapore Pte Ltd.
The registered company address is: 152 Beach Road, #21-01/04 Gateway East, Singapore 189721, Singapore

Preface

Plastics are excellent and highly useful materials to replace conventional materials including ceramic, wood, and metals. Plastics exhibit unique features of multifunctional, cheap, hygienic, light, and tunable properties. The global production and consumption of plastics have grown rapidly over years. Plastic products deliver numerous benefits to human society, and plastics are widely applied in various industries such as packaging, building and construction, automotive, electrical and electronics, and agriculture. The global plastics production is up to 390.7 Mt in 2021. The sharp growth in plastic consumption is responsible for the generation of a large amount of waste plastics. It is predicted that the worldwide production of plastic wastes would be 12,000 Mt by 2050. The improper disposal of plastic wastes associated with growing production and ineffective management pose significant burden to environment.

Microplastics have been a new dimension of waste plastics considering the negative environment impacts. The presence of small plastic debris in marine initially gained attention in the 1970s and gradually receives research interests in subsequent years. After the first mention of microplastics by Thompson's pioneering work in 2004, increasing efforts have been made to identify the abundance of microplastics and the impacts on ecosystem. Subsequently, scientists attempt to investigate the fate, contamination, and effects of microplastics on organisms, ecosystems, and Earth's natural cycles. Over the past 15 years, researchers conducted extensive researches on microplastics across the globe, resulting in tremendous advances regarding the sources, fate, and environmental impacts of microplastics. Several hundred scientific publications show that microplastics contaminate the world's oceans, including marine species at every level of the food chain, from pole to pole and from the surface to the seafloor. Microplastics are at the forefront of current environmental pollution research. Based on current knowledge, it is essential to write a book to summary the advances in the awareness of sources, fate, and solution of microplastics.

This book provides a series of review of microplastics from different perspectives. Written by interdisciplinary expert and academics, this book will attract a wide readership of students, researchers, and professionals who are interested in the fields.

Jeddah, Saudi Arabia — Anish Khan
Zhengzhou, China — Chongqing Wang
Jeddah, Saudi Arabia — Abdullah M. Asiri

Contents

About the Editors

Anish Khan is currently working as Assistant Professor in Centre of Excellence for Advanced Materials Research, Faculty of Science, King Abdulaziz University, Jeddah, Saudi Arabia. He completed Ph.D. from Aligarh Muslim University, India in 2010 and completed postdoctoral from School of Chemical Sciences, University Sains Malaysia (USM) on electroanalytical chemistry in 2010. He is working in the field of biosensor, polymers composite, organic-inorganic electrically conducting nanocomposites. He published more than 250 research articles, 80 book chapters, and 48 books in referred international publishers and more than 20 international conferences/workshop. He is member of American Nano Society, field of specialization is polymer nanocomposite/cation-exchanger/chemical sensor/microbiosensor/nanotechnology, application of nano materials in electroanalytical chemistry, material chemistry, ion-exchange chromatography and electroanalytical chemistry, dealing with the synthesis, characterization (using different analytical techniques), and derivatization of inorganic ion-exchanger by the incorporation of electrically conducting polymers. Preparation and characterization of hybrid nanocomposite materials and their applications, polymeric inorganic cation-exchange materials, electrically conducting polymeric, materials, composite material use as sensors, green chemistry by remediation of pollution, heavy metal ion-selective membrane electrode, biosensor on neurotransmitter.

Chongqing Wang is an associate professor of Zhengzhou University in Zhengzhou, China. Dr. Wang obtained his doctoral degree from Central South University, China in 2017. His academic background involves waste management, recycling of solid wastes, and wastewater treatment. Dr. Wang has experience in separation of waste plastics, microplastics management, and functional materials derived from solid wastes. He has published 80 research papers on international peer-reviewed journals, with H-index 30. Dr. Wang has obtained 15 Chinese licensed patents and 3 international licensed patents. He is the member of the International Waste Working Group, the Royal Society of Chemistry, and the Nonferrous Metals Society of China. He has been serving as Associate Editor of *Journal of the Air & Waste Management Association*, Advisory editorial board of *Clean – Soil, Air, Water*, editorial board member of *Sustainability, and Journal of Environmental Exposure Assessment*. Dr. Wang serves as Guest Editor-in-Chief of several special issue of international journals and reviewer for over 30 prestigious international journals. Dr. Wang is the Principal Investigator of eight research projects from Chinese government. E-mail: zilangwang@126.com, http://www5.zzu.edu.cn/hxgc/info/1102/3065.htm, ORCID ID: 0000-0003-2580-9263

Abdullah M. Asiri is Professor in Chemistry Department—Faculty of Science—King Abdulaziz University. Ph.D. (1995) from the University of Walls College of Cardiff, UK on Tribochromic compounds and their applications. The chairman of the Chemistry Department, King Abdulaziz University currently and also The Director of the center of Excellence for advanced Materials Research. Director of Education Affair Unit–Deanship of Community services. Member of Advisory committee for advancing materials (National Technology Plan, King Abdul Aziz City of Science and Technology, Riyadh, Saudi Arabia). Color chemistry. Synthesis of novel photochromic and thermochromic systems, synthesis of novel colorants and coloration of textiles and plastics, molecular modeling, applications of organic materials into optics such as

OEDS, high performance organic dyes and pigments. New applications of organic photochromic compounds in new novelty. Organic synthesis of heterocyclic compounds as precursor for dyes. Synthesis of polymers functionalized with organic dyes. Preparation of some coating formulations for different applications. Photodynamic thereby using organic dyes and pigments virtual labs and experimental simulations. He is member of Editorial Board of Journal of Saudi Chemical Society, Journal of King Abdul Aziz University, Pigment and Resin Technology Journal, Organic Chemistry Insights, Libertas Academica, Recent Patents on Materials Science, Bentham Science Publishers Ltd. Beside that he has professional membership of International and National Society and Professional bodies.

Interaction Between Microplastics and Pollutants

Hongru Jiang, Yingshuang Zhang, Hui Wang, and Chongqing Wang

1 Interaction Between Microplastics and Pollutants

1.1 Causes and Threats of the Interaction

The accumulation and entanglement of microplastics can cause suffocation [1], and the microplastics exposure to aquatic organisms also leads to energy depletion, fecundity, and starvation [2]. Moreover, microplastics exist in various environments as films, fragments, beads, fibers, and foams [3, 4], and their high specific surface area provides the prerequisite for carrying metal, organic pollutants, and microorganism [5]. The adsorption of pollutants on microplastics may enrich the pollutants and increase more harm to organisms [6]. Meanwhile, microplastics have been verified in the digestive tract of marine life [7], and the adsorbed pollutants may desorb and damage organs and cells [8]. The invasion of microplastics into the food web further aggregates the threat of the interaction between microplastics and pollutants. Therefore, researchers have been devoted to studying the interactions between microplastics and pollutants because they are closely related to the migration, fate, and risk of microplastics.

Microplastics in aquatic environments consist of different types, mainly including polyvinyl chloride (PVC), polystyrene (PS), polyethylene (PE), and polypropylene (PP) [9]. The different types of microplastics may show different affinities for pollutants [10]. Functional groups, specific area, and crystallinity of microplastics are regarded as the inherent factors affecting their interactions [11]. Besides, external factors in the surrounding environments, such as contact time, temperature,

H. Jiang · Y. Zhang · H. Wang
College of Chemistry and Chemical Engineering, Central South University, Changsha, China
e-mail: hongru_jiang@qq.com

C. Wang (✉)
School of Chemical Engineering, Zhengzhou University, Zhengzhou City, China

A. Khan et al. (eds.), *Microplastic sources, fate and solution*,
https://doi.org/10.1007/978-981-99-0695-6_1

dissolved organic matter, ion types, and ion strength, also affect the interaction between microplastics and pollutants [12]. After microplastics enter the natural environment, they may experience weathering, photo-degradation, mechanical friction, and microbial degradation, which indicates that surface topography may change [13]. Accordingly, compared with resin particles, natural microplastics may have different interactions with pollutants. The above factors may prompt or restrict the interactions. Chemical adsorption or physical adsorption of pollutants on microplastics surfaces is possible, and it can be justified by adsorption kinetic, adsorption isotherm, and surface characterizations. Density functional theory (DFT) as a powerful tool has also been applied to explore the interactions [14]. Besides, competitive effect and cooperative adsorption are considered to simulate various microplastics adsorption scenarios in the natural environment [15]. Researchers also further explore the desorption behavior of pollutants on microplastics, which can reveal the risks of interactions in each process [16]. Furthermore, synergistic toxic effects of microplastics and pollutants have been directly used to reveal the harm of interactions to plants and animals [17, 18]. Therefore, the research on the interaction between microplastics and pollutants has developed from adsorption characteristics and adsorption mechanism to toxic effects.

The interactions can be divided into three categories according to pollutant types, including the interaction between microplastics and metal ions, microplastics and organic pollutants, and microplastics and microorganisms. Heavy metals and metalloids may interact with microplastics, changing their environmental behaviors, bioavailability, and potential toxicity, posing ecological risks [19]. Besides, how organic pollutants interact with microplastics, such as antibiotics, is a key issue affecting ecological safety and human health [20, 21]. Furthermore, as a habitat of microorganisms, microplastics adsorbing viruses and bacteria are more biotoxic than ordinary microplastics [22]. These pollutants may interact with microplastics through complexation, electrostatic interaction, hydrophobic interaction, Van der Waals force, hydrogen bonding, π–π interaction, halogen bonding, etc. (Fig. 1) [24]. The plastic properties, external factors, and pollutants will contribute to different interaction mechanisms. This chapter introduces the interactions from adsorption characteristics to interaction mechanisms.

1.2 Interaction with Microplastics and Metal

Various reviews have observed that there are a large number of microplastics in estuary sediments, and the amount of microplastics is positively correlated with metal accumulation [25]. Natural phenomena or mining activity may lead to high concentrations of heavy metals, such as Cr, Co, Ni, Cu, Zn, Cd, Pb, Ag, and Hg [26]. In addition to the sewage from the mining area, B and As pollution has been found in the marine environment [27]. With the development of the nuclear industry, radioactive metals, such as Cs and Sr, have also appeared in the environment as emerging pollutants [28]. The existence of metals and metalloids in the natural

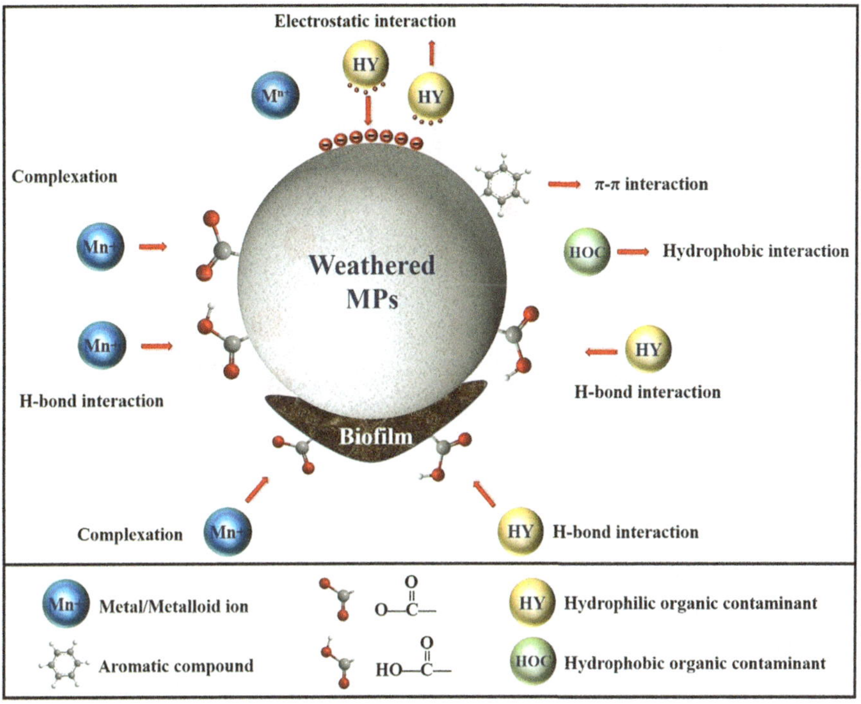

Fig. 1 The adsorption mechanism of weathered microplastics on environmental contaminants [23]

environment provides the possibility of interaction, and microplastics may have different affinities for them [29, 30].

1.2.1 Interaction with Microplastics and Heavy Metal

Haryati studied the interaction between Cr and PE microbeads in artificial seawater suspensions, and he found that PE can adsorb Cr [31]. This adsorption is relatively slow, and it takes 36 h to reach equilibrium at a concentration of 1.0 μg/mL. The excellent fitness of pseudo-first-order kinetics, Langmuir, and Freundlich model confirms chemical adsorption and monolayer adsorption.

Gao et al. suggested that aged PVC and aged PS tend to adsorb more Cu and Cd than virgin PVC and virgin PS [32]. Another article reported the adsorption behavior of three metals (Cu, Pb, and Cd) on five types of microplastics (PE, PP, PVC, PA, POM), and they found that PVC and PP tend to absorb more heavy metal than other microplastics [33].

Meanwhile, the adsorption capacity of heavy metal on PP, PVC, and PE follows the order of Pb > Cu > Cd, while PA and POM's adsorption order for heavy metals

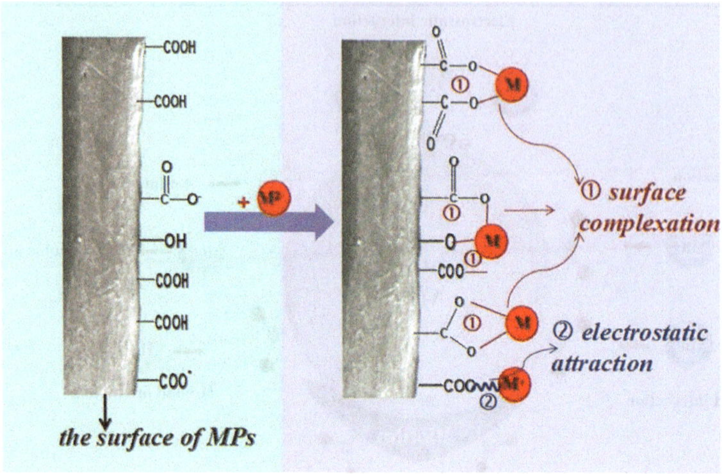

Fig. 2 The mechanism of heavy metals adsorption on microplastics [26]

is Cu > Pb > Cd. It is speculated that PVC particles have a higher adsorption capacity for metals because of the high polarity of C–Cl bond.

In the initial stage of the interaction, the metal ions are mainly adsorbed on the surface of the microplastic particles, causing the adsorption rate to increase rapidly. Furthermore, with the accumulation of metal ions, the microplastic particles are in full contact with the metal ions so that the adsorption rate slowly decreases and tends to equilibrium [30]. Heavy metals may interact with microplastics through electrostatic interaction or complexation (Fig. 2). In addition, Pb, Cu, and Cd will occupy the adsorption sites on the microplastics for competitive adsorption [33]. Holmes et al. found that the adsorption amounts of Cd, Co, Ni, and Pb sustained a net increase with decreasing salinity and increasing pH in the natural environment. However, the adsorption of Cr (VI) showed the opposite trend, and Cu almost unaffected by the above factors [34]. Other metal ions in the marine environment may compete with Cr and Cu, because microplastic surfaces become more negative with increasing pH. Moreover, long-term aging of plastic pellets and surface modification of plastic beads can affect the adsorption capacity of heavy metals to microplastics [35]. The presence of organic matter may also occupy binding sites on microplastic surfaces, so that the interaction between heavy metals and microplastics may be weakened.

1.2.2 Interaction with Microplastics and Radioactive Metal

Matthew et al. gathered microplastic samples in freshwater, estuaries, and the ocean and then characterized them with infrared spectroscopy to track changes on the plastic surface [36]. The adsorption capacity of Cs is always greater than that of Sr in all conditions. Furthermore, the Cs radiotracer has the highest adsorption in freshwater, while the Sr radiotracer has the lowest adsorption in estuaries. Cs adsorption is higher in freshwater than in other environments, owing to other cations (mostly K^+) competing for adsorption sites as salinity increases [37]. Besides, HDPE tends to adsorb more Cs and Sr than PP. Moreover, a higher percentage of Al, Si, and O in heterogeneous patches on the surface of microplastics is consistent with the accumulation in biologically facilitated deposits. Element abundance curve of the microplastic surface under estuarine conditions is consistent with clay. It is worth noting that the adsorption of the cationic alkali metal Cs and alkaline earth metal Sr is similar to that of transition metals [38]. In general, the interaction between Sr/Cs and microplastics is weak because the adsorption on microplastics is generally 2–3 orders of magnitude less than those for sediments.

1.2.3 Interaction with Microplastics and Metalloid

Wang et al. revealed that aged PVC microplastics tended to adsorb more B than other microplastics, and the chemical adsorption of B on microplastics was supported by isotherm and kinetic model [39]. Due to the conversion of boric acid ($B(OH)_3$) to borate ($B(OH)_4^-$), the interactions between boron and microplastics may depend on pH and share different adsorption mechanisms. $B(OH)_3$ can be complexed with oxygen-containing groups on microplastics surfaces, while based on electrostatic attraction, negatively charged microplastics and $B(OH)_4^-$ can be linked by cations (Fig. 3).

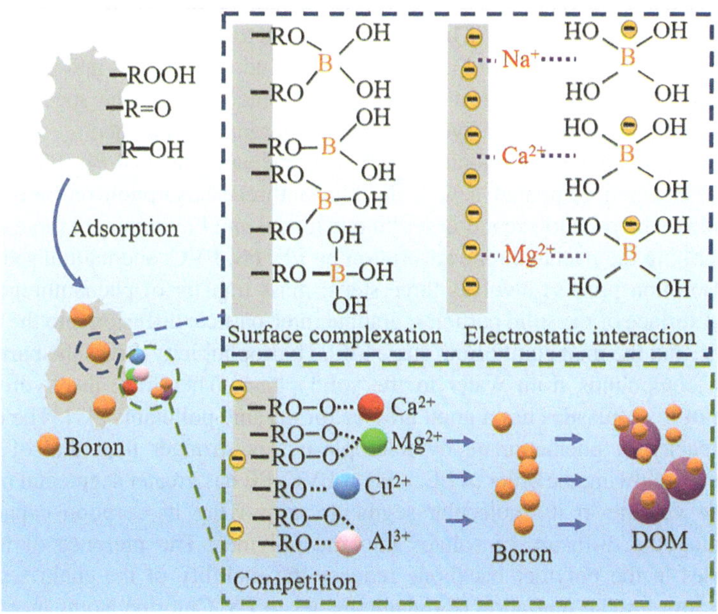

Fig. 3 The mechanism of boron adsorption on microplastics [39]

Dong et al. revealed the adsorption mechanism of As ions on polytetrafluoroethylene (PTEE) with different particle sizes, and they found that pseudo-second-order model and Langmuir model fitted adsorption kinetics and adsorption isotherm well, respectively [13]. When the pH of the solution increases, the amount of As(III) adsorbed by PTFE decreases [4]. Another vital result is that the PTFE surface can absorb the oxygen-arsenic anions, which interact with hydrogen on the hydroxyl group through electrostatic interaction. Besides, the ratio of hydroxyl and carbonyl reduces after adsorption, while As–O bonds appear [40]. The study of the adsorption of metals and microplastics should consider the changes in functional groups and the mechanisms of these changes (such as electrostatic forces and non-covalent interactions). It is worthwhile noting that NO_3^- and PO_4^{3-} restrained the adsorption of As(III) on PTFE. N, P, and As are V main group elements in the periodic table of elements [40].

1.3 Interaction with Microplastics and Organic Pollutants

1.3.1 Interaction with Microplastics and Persistent Organic Pollutants

Persistent organic pollutants (POPs) can be detected on plastic fragments in marine and freshwater environments all over the world. According to previous studies, organic pollutants can be adsorbed on the surface of microplastics and then migrated with microplastics as a carrier [41–43]. Therefore, microplastics can accumulate toxicity in a large amount and be ingested by aquatic organisms, thereby destroying the aquatic ecosystem [44]. There are many factors that affect the interaction between organic pollutants and microplastics, including the types and properties of polymers, the surface charge of microplastics, ionic strength, the specific surface area of microplastics, and the hydrophobicity of organic pollutants.

Wang et al. studied the interaction between phenanthrene and PE, PS, and PVC microplastics and compared it with the phenanthrene adsorption on natural sediments [45]. The pseudo-second-order kinetic model and Langmuir are more suitable for describing the phenanthrene adsorption on PE, PS, PVC, and natural sediments. The adsorption process involves three steps: mass transfer of phenanthrene to the external surface of the solid particles, gradual inter-particle diffusion into the interior of solids, and the final equilibrium stage [46]. Hydrophobicity drives the partition of organic compounds from water to the solid phase. Therefore, the hydrophobic surface of microplastics has a good affinity for organic pollutants [47]. The adsorption capacity of phenanthrene by microplastics is stronger than that of natural sediment, following the order of PE > PS > PVC. PE has greater segmental mobility and free volume in its molecular segments, improving its sorption capacity by facilitating the diffusion of solutes into the polymer. The presence of benzene molecules in the polymer backbone reduces the mobility of the chain segments, thereby limiting the migration of organic matter to PS. Chlorine atoms increase the density of PVC, which may reduce its free volume, thereby limiting the diffusion of organics.

Polychlorinated diphenyl ethers (PCBs), hexachlorocyclohexane (HCH), and polychlorinated biphenyls (DDE) in microplastics are closely related to the amount of local persistent organic pollutants [48, 49]. It is worth noting that compared with other microplastics, polystyrene foam has a higher concentration of polycyclic aromatic hydrocarbons (PAHs) and polychlorinated biphenyls (PCBs). Moreover, different polymer types and their properties can affect the adsorption of organic pollutants [50, 51]. Surface charge is also an important factor that affects the adsorption of organic matter, and the negatively charged microplastic surface may have a repulsive effect [50]. In addition, weathering can change the surface charge of microplastics and change the strength of adsorption [52].

Rani et al. investigated the effects of temperature, biodegradation, wave action (shaking), and water (dilution) on the interaction of hexabromocyclododecane (HBCDD) with expanded polystyrene (EPS) [53]. They found that the interaction between EPS and HBCCDs weakened, leading to further leaching of HBCCDs into the environment. EPS may be degraded or stereo-isomerized to increase the specific surface area [54].

1.3.2 Interaction with Microplastics and Persistent Organic Antibiotic

Zhang et al. found that weathered polystyrene foam has more efficient adsorption of oxytetracycline (OTC) [48]. Different pH, humic acid concentrations, and ionic strength will cause differences in adsorption of OTC by weathered polystyrene foam [55]. In addition, OTC is mainly cationic at pH <3.27, zwitterionic at pH 3.27–7.32, and anionic (OTC- and OTC 2-based) at pH 7.32. The maximum amount of adsorption can be obtained when the pH is approximately equal to five, which is close to pH_{zc}, and the electrostatic repulsion force to OTC is the lowest [56]. Ca^{2+} and Na^{+} competed for binding sites on the microplastic surface [57]. It is worthwhile mentioning that the adsorption of OTC by virgin PS is mainly determined by the cation exchange. However, the adsorption of OTC by weathered PS microplastic relies on hydrogen bonds or hydrophobic interactions [58]. Furthermore, humic acid may connect the surface of weathered PS with OTC, leading to an increase in OTC adsorption. PS can interact with aromatic structures through π–π interaction to form conjugated copolymers (Fig. 4), which may increase electrostatic attraction to zwitterions or cationic oxytetracyclines [59].

1.4 Interaction with Microplastics and Microorganisms

First, microbes can colonize the surface of microplastics and adsorb other organic or inorganic substances to form biofilms. Then, the biofilm further enriches specific microorganisms or promotes the decomposition of microplastic [60–62]. The

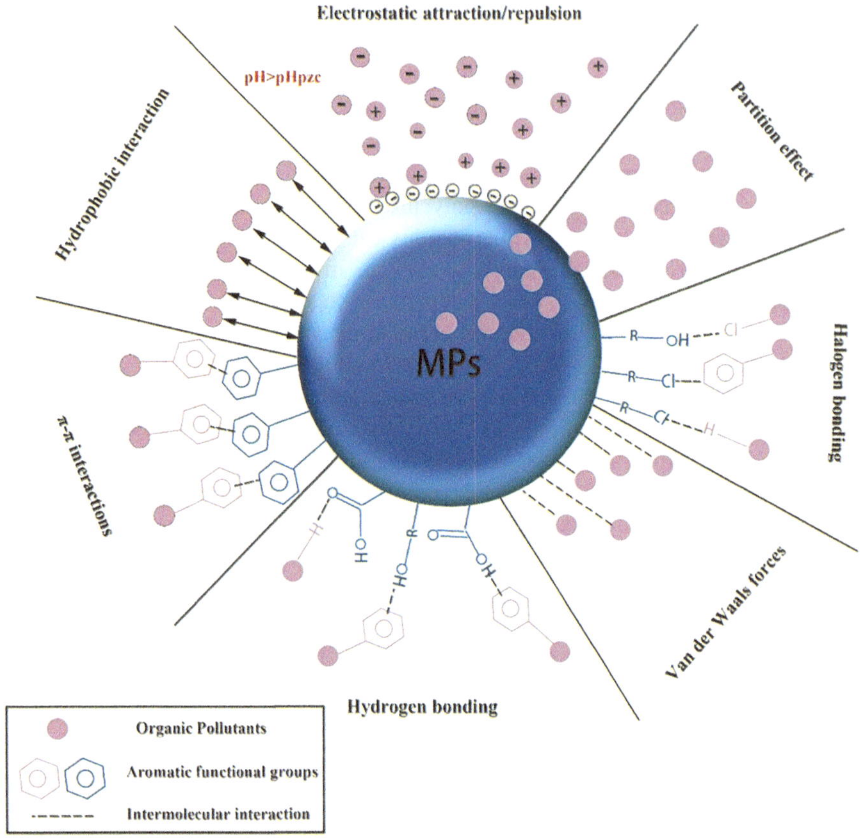

Fig. 4 The mechanism of organic pollutant adsorption on microplastics [59]

hydrophobic surface of the plastic facilitates microbial colonization and biofilm formation. Compared with the natural floating ocean substrate, plastic is more stable and has become a habitat for microorganisms and invertebrates [63]. After microorganism colonizing microplastic surface, surface structures (such as pili, curli, fimbriae, and flagella) and extracellular polysaccharides regulate surface charge and hydrophobicity of the cell walls and membranes [64]. Oberbeckmann et al. designed an exposure experiment (using PET as a sample) to investigate the effects of seasonal and spatial factors on microbial communities of marine plastic debris [65]. Surprisingly, the biofilm on the PET surface is not significantly different from the biofilm on the glass surface. There is a distinctly different community composition between the soft (e.g., litter and cardboard) and hard (e.g., tile, glass, plastic, and aluminum) substrates [66]. The availability of PET as a substrate is not vital to structure the plastic-associated biofilm communities. The great driver of microbial community composition is the surrounding environment rather than the plastic [67].

Within a few seconds after contact with water, the surface of the microplastic will absorb organic and inorganic substances to form a conditioning film. The initial conditioning film may control the colonizing community by regulating the surface characteristics [64]. Additives in microplastic will affect the organism composition [68]. Besides, compared with surface roughness and hydrophobicity of the initial substrate surface, the chemical nature of the conditioning film is more relevant for the deposition of organisms [69]. After the biofilm is formed, the floating plastic layer can be protected from ultraviolet rays in the upper water layer, thus affecting light, shear stress, oxygen, and temperature [70]. According to fluid mechanics, the formation of biofilms will significantly affect the vertical distribution of microplastics in the aquatic environment. Microplastic becomes sticky because of extracellular polysaccharides, which promote the formation of the copolymer and then change the deposition rate [64].

Long et al. studied the interaction between PS microplastics and phytoplankton using pneumococcus, dinoflagellates, and diatoms [62]. They found that when cells grow steadily, diatom exposure to PS microplastics can cause heterogeneous aggregation. Bacteria and algae can release extracellular polysaccharides and transparent exopolymeric particles to promote aggregation [71]. It is reported that PS particles affect the formation of biofilms and the composition of pelagic bacteria [61]. Microplastics can interact with algae cell membranes by inducing shielding, blocking microalgae pores, and gas exchange. After the light and air are blocked, the photosynthesis of the algae becomes weak [72]. In addition, microplastics can block the pores of microalgae and limit the exchange of energy and materials between cells and the environment. Schwab et al. have confirmed that shading is one of the main causes of the toxic effect of microplastics on microalgae [73]. However, Zhang et al. claimed that the density of chlorophyll and algae did not decrease significantly in shading experiments. The concentration of microplastics may not be enough to affect photosynthesis. Microplastic particles in the cell wall of algae can cause physical damage to the surface. Additionally, the hetero-aggregation of microplastic particles and algae also had a negative influence on the growth of microalgae [74]. The adsorption and aggregation are more likely to cause toxic effects of microplastics on algae than non-contact shading effects.

References

1. Jaafar N, Azfaralariff A, Musa SM, Mohamed M, Yusoff AH, Lazim AM (2021) Occurrence, distribution and characteristics of microplastics in gastrointestinal tract and gills of commercial marine fish from Malaysia. Sci Total Environ 799:149457
2. Shen J, Liang B, Zhang D, Li Y, Tang H, Zhong L, Xu Y (2021a) Effects of PET microplastics on the physiology of drosophila. Chemosphere 283:131289
3. Mishra S, Rath CC, Das AP (2019) Marine microfiber pollution: a review on present status and future challenges. Mar Pollut Bull 140:188–197

4. Zhang Y, Jiang H, Bian K, Wang H, Wang C (2021) A critical review of control and removal strategies for microplastics from aquatic environments. J Environ Chem Eng 9(4):105463
5. Mendoza LMR, Jones PR (2015) Characterisation of microplastics and toxic chemicals extracted from microplastic samples from the North Pacific gyre. Environ Chem 12:611–617
6. Fu Q, Tan X, Ye S, Ma L, Gu Y, Zhang P, Chen Q, Yang Y, Tang Y (2021a) Mechanism analysis of heavy metal lead captured by natural-aged microplastics. Chemosphere 270:128624
7. Browne MA, Dissanayake A, Galloway TS, Lowe DM, Thompson RC (2008) Ingested microscopic plastic translocates to the circulatory system of the mussel, Mytilus edulis (L.). Environ Sci Technol 42:5026–5031
8. Song X, Wu X, Song X, Shi C, Zhang Z (2021) Sorption and desorption of petroleum hydrocarbons on biodegradable and nondegradable microplastics. Chemosphere 273:128553
9. Xiong X, Wu C, Elser JJ, Mei Z, Hao Y (2019) Occurrence and fate of microplastic debris in middle and lower reaches of the Yangtze River–from inland to the sea. Sci Total Environ 659: 66–73
10. Jiang H, Zhang Y, Bian K, Wang H, Wang C (2022) Insight into the effect of aqueous species on microplastics removal by froth flotation: kinetics and mechanism. J Environ Chem Eng 10(3):107834
11. Mei W, Chen G, Bao J, Song M, Li Y, Luo C (2020) Interactions between microplastics and organic compounds in aquatic environments: a mini review. Sci Total Environ 736:139472
12. Lin Z, Hu Y, Yuan Y, Hu B, Wang B (2020) Comparative analysis of kinetics and mechanisms for Pb(II) sorption onto three kinds of microplastics. Ecotoxicol Environ Saf 208:111451
13. Bayo J, Olmos S, López-Castellanos J (2020) Microplastics in an urban wastewater treatment plant: the influence of physicochemical parameters and environmental factors. Chemosphere 238:124593
14. Bao Z-Z, Chen Z-F, Zhong Y, Wang G, Qi Z, Cai Z (2021) Adsorption of phenanthrene and its monohydroxy derivatives on polyvinyl chloride microplastics in aqueous solution: model fitting and mechanism analysis. Sci Total Environ 764:142889
15. Shen MC, Song B, Zeng GM, Zhang YX, Teng FY, Zhou CY (2021b) Surfactant changes lead adsorption behaviors and mechanisms on microplastics. Chem Eng J 405:12
16. Fan X, Gan R, Liu J, Xie Y, Xu D, Xiang Y, Su J, Teng Z, Hou J (2021) Adsorption and desorption behaviors of antibiotics by tire wear particles and polyethylene microplastics with or without aging processes. Sci Total Environ 771:145451
17. Dong Y, Bao Q, Gao M, Qiu W, Song Z (2022) A novel mechanism study of microplastic and as co-contamination on indica rice (Oryza sativa L.). J Hazard Mater 421:126694
18. Zhang Y, Wolosker MB, Zhao Y, Ren H, Lemos B (2020) Exposure to microplastics cause gut damage, locomotor dysfunction, epigenetic silencing, and aggravate cadmium (cd) toxicity in drosophila. Sci Total Environ 744:140979
19. Dong Y, Gao M, Qiu W, Song Z (2021) Effect of microplastics and arsenic on nutrients and microorganisms in rice rhizosphere soil. Ecotoxicol Environ Saf 211:111899
20. You H, Huang B, Cao C, Liu X, Sun X, Xiao L, Qiu J, Luo Y, Qian Q, Chen Q (2021) Adsorption–desorption behavior of methylene blue onto aged polyethylene microplastics in aqueous environments. Mar Pollut Bull 167:112287
21. Xia Y, Zhou JJ, Gong YY, Li ZJ, Zeng EY (2020) Strong influence of surfactants on virgin hydrophobic microplastics adsorbing ionic organic pollutants. Environ Pollut 265:115061
22. Lu L, Luo T, Zhao Y, Cai C, Fu Z, Jin Y (2019) Interaction between microplastics and microorganism as well as gut microbiota: a consideration on environmental animal and human health. Sci Total Environ 667:94–100
23. Duan J, Bolan N, Li Y, Ding S, Atugoda T, Vithanage M, Sarkar B, Tsang DCW, Kirkham MB (2021) Weathering of microplastics and interaction with other coexisting constituents in terrestrial and aquatic environments. Water Res 196:117011
24. Tang Y, Liu Y, Chen Y, Zhang W, Zhao J, He S, Yang C, Zhang T, Tang C, Zhang C, Yang Z (2020) A review: research progress on microplastic pollutants in aquatic environments. Sci Total Environ 766, 142572

25. Wang J, Peng J, Tan Z, Gao Y, Zhan Z, Chen Q, Cai L (2017) Microplastics in the surface sediments from the Beijiang River littoral zone: composition, abundance, surface textures and interaction with heavy metals. Chemosphere 171:248–258
26. Tang S, Lin L, Wang X, Yu A, Sun X (2021) Interfacial interactions between collected nylon microplastics and three divalent metal ions (cu(II), Ni(II), Zn(II)) in aqueous solutions. J Hazard Mater 403:123548
27. Carbery M, MacFarlane GR, O'Connor W, Afrose S, Taylor H, Palanisami T (2020) Baseline analysis of metal(loid)s on microplastics collected from the Australian shoreline using citizen science. Mar Pollut Bull 152:110914
28. Guo X, Wang J (2019) The phenomenological mass transfer kinetics model for Sr2+ sorption onto spheroids primary microplastics. Environ Pollut 250:737–745
29. Brennecke D, Duarte B, Paiva F, Caçador I, Canning-Clode J (2016) Microplastics as vector for heavy metal contamination from the marine environment. Estuar Coast Shelf Sci 178:189–195
30. Ashton K, Holmes L, Turner A (2010) Association of metals with plastic production pellets in the marine environment. Mar Pollut Bull 60:2050–2055
31. Haryati Y, Zon NF, Iskendar A, Azman S, Sarijan S, Ismail R, Nor Zurairahetty MY, Izni Syahrizal I (2018) Sorptive behaviour of chromium on polyethylene microbeads in artificial seawater. MATEC web of conferences, 250
32. Gao L, Fu D, Zhao J, Wu W, Wang Z, Su Y, Peng L (2021) Microplastics aged in various environmental media exhibited strong sorption to heavy metals in seawater. Mar Pollut Bull 169:112480
33. Gao F, Li J, Sun C, Zhang L, Jiang F, Cao W, Zheng L (2019) Study on the capability and characteristics of heavy metals enriched on microplastics in marine environment. Mar Pollut Bull 144:61–67
34. Holmes LA, Turner A, Thompson RC (2014) Interactions between trace metals and plastic production pellets under estuarine conditions. Mar Chem 167:25–32
35. Holmes LA, Turner A, Thompson RC (2012) Adsorption of trace metals to plastic resin pellets in the marine environment. Environ Pollut 160:42–48
36. Johansen MP, Prentice E, Cresswell T, Howell N (2018) Initial data on adsorption of Cs and Sr to the surfaces of microplastics with biofilm. J Environ Radioact 190-191:130–133
37. Evrard O, Laceby JP, Lepage H, Onda Y, Cerdan O, Ayrault S (2015) Radiocesium transfer from hillslopes to the Pacific Ocean after the Fukushima nuclear power plant accident: a review. J Environ Radioact 148:92–110
38. Johansen MP, Cresswell T, Davis J, Howard DL, Howell NR, Prentice E (2019) Biofilm-enhanced adsorption of strong and weak cations onto different microplastic sample types: use of spectroscopy, microscopy and radiotracer methods. Water Res 158:392–400
39. Wang H, Huang W, Zhang Y, Wang C, Jiang H (2021) Unique metalloid uptake on microplastics: the interaction between boron and microplastics in aquatic environment. Sci Total Environ 800:149668
40. Dong Y, Gao M, Song Z, Qiu W (2019) Adsorption mechanism of as(III) on polytetrafluoroethylene particles of different size. Environ Pollut 254
41. Rochman CM, Manzano C, Hentschel BT, Simonich SLM, Hoh E (2013a) Polystyrene plastic: a source and sink for polycyclic aromatic hydrocarbons in the marine environment. Environ Sci Technol 47:13976–13984
42. Rochman CM, Hoh E, Hentschel BT, Kaye S (2013b) Long-term field measurement of sorption of organic contaminants to five types of plastic pellets: implications for plastic marine debris. Environ Sci Technol 47:1646–1654
43. Andrady AL (2011) Microplastics in the marine environment. Mar Pollut Bull 62:1596–1605
44. Alimba CG, Faggio C (2019) Microplastics in the marine environment: current trends in environmental pollution and mechanisms of toxicological profile. Environ Toxicol Pharmacol 68:61–74

45. Wang W, Wang J (2018) Different partition of polycyclic aromatic hydrocarbon on environmental particulates in freshwater: microplastics in comparison to natural sediment. Ecotoxicol Environ Saf 147:648–655
46. Cheung WH, Szeto YS, Mckay G (2007) Intraparticle diffusion processes during acid dye adsorption onto chitosan. Bioresour Technol 98:2897–2904
47. Wang J, Tan Z, Peng J, Qiu Q, Li M (2016) The behaviors of microplastics in the marine environment. Mar Environ Res 113:7–17
48. Zhang H, Wang J, Zhou B, Zhou Y, Dai Z, Zhou Q, Chriestie P, Luo Y (2018) Enhanced adsorption of oxytetracycline to weathered microplastic polystyrene: kinetics, isotherms and influencing factors. Environ Pollut 243:1550–1557
49. Ogata Y, Takada H, Mizukawa K, Hirai H, Iwasa S, Endo S, Mato Y, Saha M, Okuda K, Nakashima A (2009) International pellet watch: global monitoring of persistent organic pollutants (POPs) in coastal waters. 1. Initial phase data on PCBs, DDTs, and HCHs. Mar Pollut Bull 58:1437–1446
50. Baick IH, Yang WJ, Ahn YG, Song KH, Choi KY (2015) Structure and properties of ultra-high molecular weight bisphenol a polycarbonate synthesized by solid-state polymerization in amorphous microlayers. J Appl Polym Sci 132:n/a-n/a
51. Guo X, Wang X, Zhou X, Kong X, Tao S, Xing B (2012) Sorption of four hydrophobic organic compounds by three chemically distinct polymers: role of chemical and physical composition. Environ Sci Technol 46:7252–7259
52. Fotopoulou KN, Karapanagioti HK (2015) Surface properties of beached plastics. Environ Sci Pollut Res 22:11022–11032
53. Rani M, Shim WJ, Jang M, Han GM, Hong SH (2017) Releasing of hexabromocyclododecanes from expanded polystyrenes in seawater-field and laboratory experiments. Chemosphere 185: 798–805
54. Zhao Y-Y, Zhang X-H, Sojinu OS (2010) Thermodynamics and photochemical properties of α, β, and γ-hexabromocyclododecanes: a theoretical study. Chemosphere 80:150–156
55. Liang X, Chen B, Nie X, Shi Z, Huang X, Li X (2013) The distribution and partitioning of common antibiotics in water and sediment of the Pearl River estuary, South China. Chemosphere 92:1410–1416
56. Sun H, Shi X, Mao J, Zhu D (2010) Tetracycline sorption to coal and soil humic acids: an examination of humic structural heterogeneity. Environ Toxicol Chem 29:1934–1942
57. Li J, Zhang K, Zhang H (2018) Adsorption of antibiotics on microplastics. Environ Pollut 237: 460–467
58. Ocampo-Pérez R, Leyva-Ramos R, Rivera-Utrilla J, Flores-Cano J, Sánchez-Polo M (2015) Modeling adsorption rate of tetracyclines on activated carbons from aqueous phase. Chem Eng Res Des 104:579–588
59. Fu L, Li J, Wang G, Luan Y, Dai W (2021b) Adsorption behavior of organic pollutants on microplastics. Ecotoxicol Environ Saf 217:112207
60. Oberbeckmann S, Kreikemeyer B, Labrenz M (2018) Environmental factors support the formation of specific bacterial assemblages on microplastics. Front Microbiol 8:2709
61. Kesy K, Oberbeckmann S, Muller F, Labrenz M (2016) Polystyrene influences bacterial assemblages in Arenicola marina-populated aquatic environments in vitro. Environ Pollut 219:219–227
62. Long M, Paul-Pont I, Hegaret H, Moriceau B, Lambert C, Huvet A, Soudant P (2017) Interactions between polystyrene microplastics and marine phytoplankton lead to species-specific hetero-aggregation. Environ Pollut 228:454–463
63. Virsek MK, Lovsin MN, Koren S, Krzan A, Peterlin M (2017) Microplastics as a vector for the transport of the bacterial fish pathogen species Aeromonas salmonicida. Mar Pollut Bull 125: 301–309
64. Rummel CD, Jahnke A, Gorokhova E, Kühnel D, Schmitt-Jansen M (2017) Impacts of biofilm formation on the fate and potential effects of microplastic in the aquatic environment. Environ Sci Technol Lett 4:258–267

65. Sonja O, Loeder MG, Gerdts G, Osborn A (2014) Spatial and seasonal variation in diversity and structure of microbial biofilms on marine plastics in northern European waters. FEMS Microbiol Ecol 90(2):478–492
66. Hoellein T, Rojas M, Pink A, Gasior J, Kelly J (2014) Anthropogenic litter in urban freshwater ecosystems: distribution and microbial interactions. PLoS One 9(6):e98485
67. Lee OO, Wang Y, Tian R, Zhang W, Shek CS, Bougouffa S, Al-Suwailem A, Batang ZB, Xu W, Wang GC (2014) In situ environment rather than substrate type dictates microbial community structure of biofilms in a cold seep system. Sci Rep 4:3587
68. Restrepo-Flórez J-M, Bassi A, Thompson MR (2014) Microbial degradation and deterioration of polyethylene–a review. Int Biodeterior Biodegradation 88:83–90
69. Lorite GS, Rodrigues CM, de Souza AA, Kranz C, Mizaikoff B, Cotta MA (2011) The role of conditioning film formation and surface chemical changes on Xylella fastidiosa adhesion and biofilm evolution. J Colloid Interface Sci 359:289–295
70. Long M, Moriceau B, Gallinari M, Lambert C, Huvet A, Raffray J, Soudant P (2015) Interactions between microplastics and phytoplankton aggregates: impact on their respective fates. Mar Chem 175:39–46
71. Passow U (2002) Transparent exopolymer particles (TEP) in aquatic environments. Prog Oceanogr 55:287–333
72. Bhattacharya P, Lin S, Turner JP, Ke PC (2010) Physical adsorption of charged plastic nanoparticles affects algal photosynthesis. J Phys Chem C 114:16556–16561
73. Schwab F, Bucheli TD, Lukhele LP, Magrez A, Nowack B, Sigg L, Knauer K (2011) Are carbon nanotube effects on green algae caused by shading and agglomeration? Environ Sci Technol 45:6136–6144
74. Zhang C, Chen X, Wang J, Tan L (2017) Toxic effects of microplastic on marine microalgae Skeletonema costatum: interactions between microplastic and algae. Environ Pollut 220:1282–1288

Microplastics in the Freshwater and Earthbound Conditions: Prevalence, Destinies, Impacts, and Supportable Arrangements

Elena Gregoris, Beatrice Rosso, Marco Roman, and Fabiana Corami

1 Introduction: Microplastics in Freshwater

An increasing number of researches have been published on MPs' different freshwater ecosystems worldwide, including lakes, rivers, estuaries, and wetlands in both waters and sediments. Although the literature on plastics in freshwater is growing, data are globally disjointed about the evidence of concentrations of MPs and impacts on environmental freshwater matrices [1]. Besides, the heterogeneity of sampling protocols, methodologies, techniques, and measurement units among different researches makes it challenging to compare results and produce a complete picture of the existing state of understanding [2]. Another critical factor is that MPs < 100 μm (SMPs) are often overlooked; less than 20% of reported studies on freshwater environments reported MPs below 20 μm. The sampling and analytical procedures are strictly connected to detecting plastic particles sizes >100 μm. Consequently, most of the reported studies could have undervalued the concentration of SMPs. Thus, improvement in detection technologies and pre-treatments methods could modify the final mean amount of MPs in freshwater environments [3].

Freshwater systems are considered an important source and critical pathway for MP pollution in coastal and marine environments, especially for the point sources close to areas with a high density of population and civilization [4–6]. Research in freshwater environments has detected MPs in Europe, North America, Africa, Asia,

E. Gregoris · B. Rosso · F. Corami (✉)
Institute of Polar Sciences, National Research Council – CNR ISP, Venezia, Italy

Department of Environmental Sciences, Informatics and Statistics, DAIS, University of Ca' Foscari, Venezia, Italy
e-mail: fabiana.corami@cnr.it; f_corami@unive.it

M. Roman
Department of Environmental Sciences, Informatics and Statistics, DAIS, University of Ca' Foscari, Venezia, Italy

A. Khan et al. (eds.), *Microplastic sources, fate and solution*,
https://doi.org/10.1007/978-981-99-0695-6_2

and Antarctica. The USA has performed the most research on MPs of all freshwater microplastics studies, followed by China [7]. Regarding water, the most polluted areas are Southeast Asia and Europe, while sediments are also highly contaminated in North Africa and North America [1]. Thus, MPs are widespread pollutants not only in marine systems but also in freshwater systems. Studies on MPs in freshwater systems are mainly conducted in rivers, lakes, and estuaries, while new research has focused on wetlands, stormwater ponds, and reservoirs. However, there is limited knowledge about MPs in these freshwater environments compared to rivers and lakes [2, 8].

2 Occurrence and Evidence of MPs in Freshwaters

2.1 *Occurrence and Evidence of MPs in Rivers*

Microplastic pollution is critical in estuaries worldwide, indicating that terrestrial river contribution is a crucial source of microplastics (Figs. 1 and 2). It is estimated that 1.15–2.41 million tonnes of plastic waste enter the ocean through rivers annually [9]; a total of 70–80% of MPs in the marine environment are introduced through rivers [10]. The top 20 polluting rivers, primarily located in Asia, account for 67% of the global total.

MP occurrence in riverine waters could vary spatially and temporally. The first report on the abundance and composition of plastic litter in a large river was in the Austrian Danube ([11]; $n = 17{,}349$; mean ± S.D: 316.8 ± 4664.6 items per 1000 m^3). Then, in Europe, [12] analyzed River Trent and River Leen, UK; they discovered that MPs were presented in low concentrations (max 0.4 particles L^{-1}) at all freshwater collection sites, including upstream of urban areas. 77% of particles had the largest dimension (65–500 μm). A higher number of MPs were detected by [13], where an average of 100 particles L^{-1} and a maximum of 187 particle L^{-1} were discovered in the water phase of the urban canal of Amsterdam.

Also, different studies in China reveal a high abundance of MPs in estuaries located in populated urban areas; however, the current evidence of MP pollution in this country is affected mainly by ineffective management and lack of applicable regulation [14]. Research at the Chinese Yangtze River mouth showed significantly higher MP concentration (average of 4137 particles m^{-3}) than most of the sampled river worldwide [15], with an annual input of 0.33 million tonnes of plastic discharged into the East China Sea [9]. Also, the Pearl River in China was investigated. A sampling of MPs on 25 beaches (average of 520 ± 688 items m^{-2}) indicated that this river is a significant source of plastic pollution for the region, and expanded polystyrene debris represented a vast majority of the samples [16]. In the Renhuai basin of the Chishui River in China, the abundance of MPs ranges from 1.77 to 14.33 items L^{-1}. The main shape was fibrous (59.4%), and the size was mainly 500–1000 μm (63.9%). Hence, the study underlined that long MPs could be subjected to degradation, producing a high number of small and medium-sized MPs

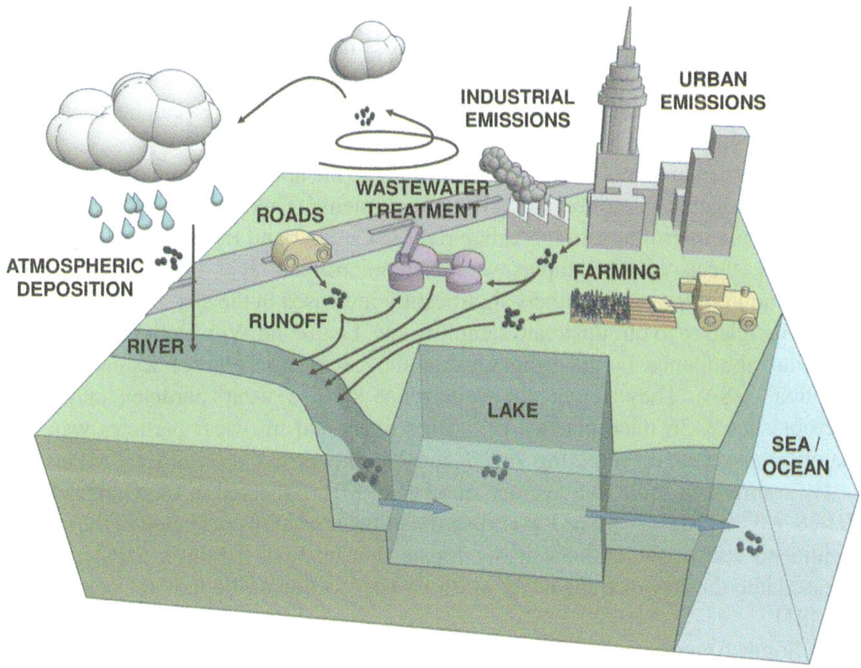

Fig. 1 Schematic representation of pathways and routes of microplastics in different environmental matrices. The freshwaters are a primary pathway for the transport of microplastics toward other environmental compartments

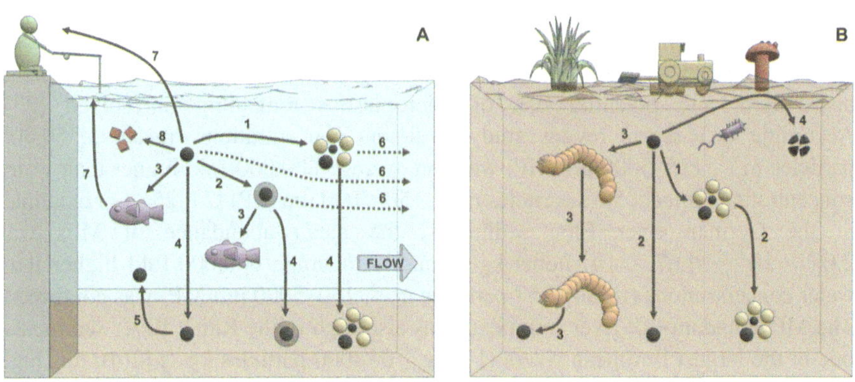

Fig. 2 Schematic representation of the main processes playing a role in the transport of MPs in (**a**) freshwater and (**b**) soil. In freshwater, the represented processes are 1. aggregation; 2. biofilm formation; 3. uptake by aquatic biota; 4. sedimentation; 5. resuspension; 6. transport to the sea/ocean; 7. uptake by humans (water withdrawal or fish consumption); 8. additives release. In soil, the represented processes are 1. aggregation; 2. vertical transport to deeper layers; 3. uptake, transport, and release by terrestrial organisms; 4. degradation mediated by fungi or bacteria

[5]. More substantial evidence of MPs was also discovered in the Lijiang River in Guangxi, Southwest China [17]. The average number of MPs was 67.5 ± 65.6 items m^{-3}. The most abundant site was located close to the outlet of WWTPs. Even though MPs in wastewater can be effectively removed, many MPs can be discharged to the receiving waters [18].

The presence of WWTPs can also contribute to the abundance and distribution of also anthropogenic microfibers in river environments. Therefore, [19]) investigated anthropogenic microfibers in this urban river in Cumberland River in Nashville, TN, USA. The mean of all samples was 27.9 microfibers L^{-1} and that the mean concentration of these microfibers significantly increased in the city and downstream of the city relative to our upstream sampling site. Leads and Weinstein [20] observed a similar abundance in the tributaries of the Charleston Harbor Estuary, South Carolina, USA. They found the occurrence of tire wear particles and other microplastics 3–36 microplastics L^{-1}. Blue fibers and tire wear particles were the majority observed, representing 26.2% and 17.1%, respectively, of total MPs.

By contrast, in India, the average number of MPs collected in the Ganges River (0.038 ± 0.004 MP L^{-1}) was lower than the number of items collected in the studies mentioned above; the authors also estimate that up to 1–3 billion MPs could be released into the Bay of Bengal (the north-eastern portion of the Indian Ocean) every day [21].

In South America, at the Elqui, Maipo, Biobio, and Maule Rivers, the research showed a consistent relation between the numerical number of MPs (range 0.05–0.74 particles m^{-3}) and the presence of litter deposition on riversides [22].

MP abundances differed between the water and sediment phases in river environments. Hence, sediments seem to be a long-term sink for MPs. However, to date, very few studies have focused on the presence of microplastics in freshwater sediment. Our understanding of microplastic pollution in freshwater sediments is relatively limited, with some relevant issues regarding the quantity, source, spatial distribution, and potential risk of microplastics remaining unresolved [23]. According to a recent review study, sediment concentrations (mean: 3,350,000 particles m^{-3}, 125–5000 μm MP) were on average 600,000-fold higher than water concentrations (mean: 5.57 particles m^{-3}, 150–5000 μm MP) [24, 25]. For example, in the German river Elbe sediments, the mean abundance of MPs was $2.08 \times 10^3 \pm 4.67 \times 10^3$ items kg^{-1}, and an average 600,000-fold higher than water concentrations (mean: 5.57 particles m^{-3}, 150–5000 μm MP) was confirmed. The MP abundance of river sediments was also high in the Rhine River sediments, one of the largest European rivers (1 g kg^{-1} or 4000 particles kg^{-1}; [26]); this high abundance was also confirmed by Mani and Burkhardt-Holm [27]. Different benthic midstream sediment samples were collected, and the concentration of MPs ranged between 0.26 ± 0.01 and $11.07 \pm 0.6 \times 10^3$ MP kg^{-1} with a decrease in sediment depth. Eighteen polymers were identified (size range of 11–500 μm). Nevertheless, a low abundance in sediments was observed in the Tisza River in Central Europe (average abundance 3177 ± 1970 items kg^{-1}; [28]), in the urban river in Scotland, UK (161–432 MPs kg^{-1}; [29]), and in a river catchment of central Spain (range of 0–2630 MPs kg^{-1} dw).

2.2 Occurrence and Evidence of MPs in Lakes

Since lakes are closed water bodies surrounded by land, their water movement is relatively low than those in rivers; hence, MPs have a higher potential for settling in lakes. MP occurrence in lakes was strictly connected to intensive anthropogenic activities and population density [30]. Along the beaches of Lake Huron, Canada, a total of 3209 plastic pieces were found, and a high number of plastic pellets that consist of 94% of the total plastic debris collected were observed, especially proximal to human activities near an industrial sector along the margin of the lake [31]. Also, Ballent et al. [32] observed that the abundance of MPs in Canadian Lake Ontario sediments (with an average of ~760 particles kg^{-1} dw) might be attributed to the high population density and industrial zones in the watersheds draining into this region. The ubiquitous presence of MPs in sediments was reported in other lakes of the USA, such as in Lake Michigan and Lake Erie; an average abundance of 65.2 items kg^{-1}, whose size was >355 μm, was reported in the sediments of Lake Michigan, while an average abundance of 431 items kg^{-1} was observed in the sediments of Lake Erie [33]. The sources include the surroundings with high-density population regions, agricultural land cover, and upstream areas. Similar sources were evidenced for Lake Mead, USA [34]; the abundance of MPs in surface water ranged from 0.44 to 9.7 items m^{-3}, while that in sediments ranged from 87.5 to 1010 items kg^{-1} dw. The relevant abundance of MPs in lakes was due to heavy anthropogenic impacts in the surroundings.

Studies on European lakes showed how MP pollution is extended worldwide. In Central Italy [35], MPs accounted for an abundance of 2.68–3.36 items m^{-3} in the surface waters of Lake Chiusi, while the abundance ranged from 0.82 to 4.42 items m^{-3} in the surface waters of Lake Bolsena. Sediments of these two lakes showed significantly higher MP abundance than waters (234 items kg^{-1} dw and 112 items kg^{-1} dw, respectively). The catchment area and heavy winds in the areas of the lakes might affect MP abundance in the surface water layer. MPs also occurred in the sediments of Vesijärvi lake, Finland (average of 395.8 ± 90.7 MPs kg^{-1}; [36]), and in the Caspian Sea, the largest lake in the world (from 25 to 330 items kg^{-1}). In the latter basin, the presence of MPs was related to fishing and tourism activities [37]. Also, Lake Onego in Russia, the second largest lake in Europe, showed a relevant abundance of MPs in the sediments (2188.7 ± 1164.4 items kg^{-1} dw; [38]).

A high concentration of MP pollution was also detected in China. Domestic sewage and fishing activities might be the primary sources of MPs in Poyang Lake, the largest freshwater lake in China; 5–34 items L^{-1} for surface waters and 54–506 items kg^{-1} for sediments were observed [39]. In addition to human activities, MP pollution of Poyang Lake might also be influenced by hydrodynamic and topographic conditions. Although the narrow channel from the north could increase suspended particles, the hydrodynamic features of the southern region may dilute the amount of plastic pollution and diminish the impacts of land-based plastic debris. The influence of the current lake correlated to the evidence of MPs in lakes was also investigated by Xiong et al. [40] in another big lake of China, the Qinghai Lake. MPs were detected with abundance varying from 0.05 × 105 to 7.58 × 105 items km^{-2} in the lake surface water, from 0.03 × 105 to 0.31 × 105 items km^{-2} in the inflowing rivers, and from 50 to 1292 items m^{-2} in the lakeshore sediment. The lake circulation might accumulate MPs in the middle part of the lake, caused by the transport of lake current. Different concentrations of MPs were also found in Wuliangsuhai Lake, where the abundance of MPs (0–6 cm) in surface sediment ranged from 16.5 to 72.4 particles 100 g^{-1} dw [41]. In Lake Ulansuhai, the abundance of MPs ranged from 24 ± 7 to 14 ± 3 items kg^{-1} [42], while in East Dongting Lake, MPs ranged from 180 to 693 items kg^{-1} [43]. In addition, Yin et al. [43] showed that the abundance of microplastics in the sediments of rural areas is higher than that in urban areas. Hence, in urban areas, wastewater management helps to prevent the pollution of MPs in receiving water bodies, while in rural areas, there are often only poor treatment facilities, and the wastewater is discharged directly into the environment.

MPs have also been reported in remote lakes with restricted human activities. For example, Velasco et al. [44] found in a Remote Mountain Lake in Switzerland an average of 2.6 MPs and 4.4 fibers L^{-1} in the water column and 514 fibers and 33 microplastics kg^{-1} in sediments. The evidence of MPs in remote lakes is also confirmed by Zhang et al. [45]. They collected different lakeshore sediments from remote lakes in Tibet plateau, China, and the average of MPs found was from 8 ± 14 to 563 ± 1219 items m^{-2}, explaining that freshwater input might have an outstanding contribution. A high average abundance of MPs in remote lake sediments (20,264 particles km^{-2}) was also shown in Lake Hovsgol, Mongolia. The authors underlined that the results might be partially explained by its long residence time and small surface area of the remote lake [46].

2.3 *Occurrence and Evidence of MPs in Stormwater Ponds and Reservoir*

Most studies focus on the rivers and lakes worldwide. However, there is limited understanding of MP pollution in other freshwater systems, such as groundwater, ponds, and reservoirs. Urban runoff, highways, and agriculture can contribute to MP pollution in surface waters; hence ponds and wetlands are considered essential reservoirs for accumulating MPs. Townsend et al. [47] presented one of the first studies highlighting MP pollution (<1 mm) in Australian urban wetlands and emphasizing the ubiquitous microplastic pollution in the freshwater environment. The average MP abundance in sediments was around 46 items kg^{-1} dw. The concentration increased in wetlands, reservoirs, and catchments near industrial areas and decreased in catchments with higher residential densities. Accumulation of MPs in reservoirs can also be affected by WWTPs, fishing, and agricultural activities [48]. In Danjiangkou Reservoir, the second largest reservoir in China, the mean value of MP pollution was 7205 items m^{-3} for water and 1818 items kg^{-1}

for sediments [49]; hence, agricultural cultivation and hydro-fluctuation can be a significant source of MPs in reservoirs. Also, stormwater runoff has received little attention but has been documented as a source that can release MPs to receiving waters. The first assessment of MPs in the sediments of stormwater retention ponds in Denmark was investigated by [50]. The highest (127,986 items kg^{-1}; 28,732 μg kg^{-1}) accumulation of MPs was found in the sediments of ponds near industrial activities. The stormwater pond's MPs were also highly concentrated in sediments from Denmark, reaching 0.4 g kg^{-1}, corresponding to nearly 106 item kg^{-1} [51]. Despite the attention paid to MP contamination, recent research highlighted the massive presence of synthetic rubber MPs from tire particles from roads and highways that can reach the environmental water bodies. For example, between 15% and 38% of MPs in the sediments from Australian stormwater floating treatment wetland came from car tires. The average of 595 ± 120 and 320 ± 42 MPs items kg^{-1} dw in inlet and outlet were analyzed [52]. Further analysis on the evidence of tire wear particles needs to be investigated, especially for freshwater water bodies [53].

3 Occurrence of MPs in Earthbound Condition on Freshwater Environments

Microplastics can enter freshwater bodies from various sources by stormwater runoff, wastewater treatment plant (WWTP) effluent, agricultural/urban activities, and atmospheric deposition. MPs present in the soil and terrestrial environments can be transferred into the river catchments and contaminate the freshwater environment. Thus, the soil is considered a sink for MP accumulation and a source for the surrounding environments [54]. Terrestrial hotspots for MP pollution from urban areas can contain the highest number of MPs, possibly due to their closeness to the sites within the watershed where plastics are released. Therefore, soil samples from areas closer to industrial emission sites are characterized by a high concentration of MPs. This was confirmed by Rafique et al. [55] who found an average concentration of 5780 ± 3251 MPs kg^{-1} in soils of industrial areas in Pakistan. Industrial soils were also analyzed in Sydney, Australia, and the results indicated that the occurrence of MPs varied in the range of 300–67,500 mg kg^{-1} [56]. Other studies were conducted in Germany, where the soil samples contained 24,000 MPs kg^{-1} [57] and in China with an average concentration of 18,760 MPs kg^{-1} [58].

Agricultural soils typically receive MPs from plastic employed for weed control, wastewater biosolids and compost, polymer-based fertilizers, and atmospheric deposition [3, 59]. Evidence of MPs between 80% and 90% in agricultural soils is presented in sludge derived from urban water treatments, without an efficiency blocked by WWTPs. Hence, the resulting biosolids potentially contaminated are often used as fertilizer to soils [60]. Eastern Spanish agricultural soils with the application of sludge, deriving from different WWTPs, had an average of 5190

MPs kg^{-1} [61]. Also, synthetic fabric fibers could be markers of the use of wastewater sludge in agricultural soil [62]. Most plastic fibers were also analyzed in different agricultural soils from sewage sludge disposal in Chile [63]; MPs ranged from 18 to 41 particles g^{-1}, with a median of 34 particles g^{-1} in 31 agricultural fields with different sludge application records. Compost samples of rural domestic waste in China were investigated, and an average MP abundance of 2400 ± 358 items kg^{-1} dw was found [64]. While Weithmann et al. [65] detected 14–895 items kg^{-1} MPs in the organic fertilizer converted from biosolid, Bläsing and Amelung [57] observed an average of 2.38–180 mg kg^{-1} of MPs in the organic fertilizer. Hence, fertilizers may be a source of MPs for soils.

Also, landfills are receptacles for MP waste derived from industry and households, which can be accumulated; however, evidence on MP abundance in landfill systems is needed. Kilponen [66] showed the first evidence on this topic in an old-closed landfill in Finland. They reported MP concentrations up to 1.1 items L^{-1} in leachate. Another study [67] revealed that in Shanghai, the average abundances of MPs in leachate and litter were 8 (±3) items L^{-1} and 62 (±23) items g^{-1}, respectively. He et al. [68] underlined that 99.36% of MPs in landfills are derived from the fragmentation of plastic waste, and they found a concentration of MPs ranging from 0.42 to 24.58 items L^{-1} in the municipal solid waste landfill in China.

However, some authors confirmed that WWTPs mostly remove MPs transferred to landfills; for instance, Ziajahromi et al. [52] found that only 0.2–8% of these MPs were detected in effluent from WWTPs, and 8–16% were found in sludge, ranging from 15.9 to 56.5 particles g^{-1} dw. A similar result was obtained by Lares et al. [69]: 98.3% of MPs were removed during conventional WWTP processes. In this study, MP concentrations varied between 0.1 and 124.7 items L^{-1} in water samples and between 8.2 and 301.4 items g^{-1} dw in sludge samples. It should be noted that small MPs (<100 μm) may.

4 Fate of MPs

4.1 Fate of MPs in Freshwater

Transport of particles in freshwaters differs a lot among the various bodies, depending on the system properties. In rivers, the high flow velocity of water maximizes the transport of MPs to the sea, whereas in lakes that have a flow velocity very low or absent, the sedimentary processes are enhanced. Tibbetts et al. [70] quantified microplastics in the River Thames (UK) and demonstrated that MP concentration is related to the population density and the presence of lakes and meanders, where the flow velocity is reduced. Therefore, biological and chemical processes are usually more important in lakes, acting as a sink of MPs and pollutants. MP transport also depends on the waterbody geometry, depth, weather conditions (e. g., effect of wind), and other waterbody characteristics [71, 72]. Furthermore, water withdrawal for various human uses could influence the fate of plastics, as it has been demonstrated for nutrients [73].

Moreover, the transport of microplastics strongly depends on their density and contact angle. The particle density of PP and PE is lower than water, indicating that they will be mainly floating in the water, with a slight tendency to settle to the sediment. Conversely, PVC and PET will be quickly removed from water by sedimentation due to their density being higher than water. After settling, particles can reenter the water column by resuspension [3]. The hydrophobic interaction between particles is the cause of the aggregation behaviors of MPs, which can undergo various processes, namely aggregation and biofouling. Aggregates can be formed by coagulation of MPs or with organic/inorganic suspended particulate; they usually have a higher density than the original MPs, thus influencing their fate: this effect could enable light MP particles to reach the benthic zones and to sink into sediments. It has also been demonstrated that the formation of biofilms of microorganisms on the surface of MPs can change some plastic physical properties, such as density and hydrophobicity [74], affecting their transport.

4.2 Fate of MPs in Soils

Microplastics in soils can originate from the fragmentation of plastic waste and items, from atmospheric deposition, surface runoff, or irrigation using contaminated wastewater [75]. Once released into soils, MPs can persist and accumulate, modifying the physicochemical characteristics of soils and affecting the terrestrial ecosystems (discussed in detail in Sect. 5).

Organisms play an essential role in determining the fate of MPs. Fungi and bacteria in the topsoil can slowly decompose plastics, using plastics as a carbon source [76], and fungi are the principal colonizers during the biodegradation of different forms of plastics. Earthworms can act as carriers of MPs, especially small MPs (<100 μm), transporting them to the deeper layers, with the smallest particles moving proportionally the most [77]. The exact transport mechanisms are still unknown, but they could include adherence of the MPs externally to the earthworm, movements of the earthworm, and the ingestion/egestion mechanism, which was also documented by Huerta Lwanga et al. [78]. The vertical transport of MPs has significant implications: since the speed of degradation of organic material is generally smaller in deep soil, MP persistence could be enhanced when these particles reach greater depths [79].

On the other hand, MPs in the topsoil are subjected to direct exposure to UV radiation from sunlight and high-temperature fluctuations [80], which can potentially be a degradation environment for MPs [68, 81], leading to the formation of smaller pieces with a different fate [79]. Other potential mechanisms have been proposed for MPs entering deep soil, such as agricultural cultivation, soil crack, and bioturbation by plant roots [5]. MP movements in soils depend on soil characteristics (e.g., porosity), weather conditions (e.g., dry/wet cycles, [82]), and properties of the particles themselves (e.g., size, shape, density). Generally, the smaller the particles, the more they move downward, passing through soil pores [79]. Low-density MPs might slowly leach downward [82]. Currently, no studies have clarified in depth the effect of shape on MP migration in soils and the effects of MPs' type and surface structure on their migration [5]. Besides sedimentation and interaction with biota, microplastics could also be incorporated into soil aggregates with organic matter, microbes, and primary soil particles, in a dynamic process, with aggregates being formed and disintegrating [77]. The formation of aggregates depends on microplastics properties (e.g., particle size and surface structure) and environmental factors (e.g., pore size, ionic strength, and saturation) and could limit the transport of MPs in soil [5].

5 Impacts of Microplastics

MPs can influence the physicochemical characteristics of water, causing impacts on aquatic biota. Various studies have demonstrated that a high concentration of MPs can affect algae growth [83, 84], blocking the light transport or adsorbing on the algae surface. MPs can also modify the chemistry of water, leaching additives, and plasticizers in the waterbody or adsorbing chemicals from the surrounding environment. MPs have the potential of acting as a vector of pollutants and substrates for microorganisms [85, 86]; they can also interact with aquatic biota by entanglement or being ingested. MPs in soils can produce physicochemical changes in soil texture and structure, which have consequences for water cycling and ecosystem functioning in the terrestrial environment [5]. Changes in the hydrologic properties of soils could influence soil microbial biodiversity, which can alter soil quality [87]. Besides, MP surfaces are known to interact with hydrophobic compounds, which have a role in soil hydrophobicity and soil aggregate stability, with direct potential consequences for soil erosion and biogeochemical cycles. However, further research is required to clarify the extent to which microplastic pollution could affect soil chemistry, texture, structure, and function [88].

The most concerning environmental effect of the occurrences of MPs in freshwater and soil is the impact on biota and human health. Laboratory and field studies have demonstrated that species of freshwater invertebrates and fish can ingest microplastics [89], with substantially no difference across habitats, feeding guilds, and trophic levels. The accumulation of MPs in the terrestrial food web is very little studied. Microplastics have been detected in terrestrial birds [90]. However, an increasing number of laboratory studies are now focusing on the potential impact of MPs on terrestrial animals and plants, evidencing adverse effects, such as physical changes and behavioral effects [78, 87, 91–93].

6 Microplastic Effects on Biota

6.1 Physical Effects

Possible physical effects due to the presence of MPs are entanglement and ingestion. Entanglement could cause several impacts on aquatic species, including turtles, sea birds, marine mammals, fish, crustaceans, and squid [94]. Once caught, animals unable to free themselves can drown, suffocate, starve because of limited mobility, or be easy prey for predators. The proportion of freshwater species at risk of entanglement seems lower than marine species but still substantial [95].

The pathway of entry of MPs into the organism includes direct and indirect ingestion. Invertebrates ingest MPs according to the size of their mouthparts; by the ingestion of water and soil debris, MPs can be ingested directly and enter the trophic web. Once in the trophic web, MPs can be accumulated along with the different levels, which may also be magnified. The magnification throughout the food web was calculated as a factor between 22 and 160 in seals [96]. The first evidence of trophic transfer of microplastics was demonstrated in a terrestrial food chain [97]. Ingestion of microplastics can cause several effects on aquatic species, mainly due to the digestive tract's blockages and the accumulation of plastics in the digestive cavities, which can cause a false satiation sensation and a reduced dietary intake by organisms. The resulting physical deterioration can lead to reduced reproductive fitness, diminished predator avoidance, and impairment of feeding ability. The studies on freshwater fauna are limited compared to marine species. Adverse effects were observed in freshwater crustaceans [98–100] and fish [101]. However, some works showed no effects on feeding activity, energy reserves, molt, or mortality [102, 103], due to ingestion of MPs. This contradiction is maybe due to the different exposure conditions or target species selected in the works.

Regarding terrestrial biota, the limited number of studies focusing on the toxicity of MPs does not discriminate the toxicity mechanism; thus, we cannot distinguish whether the observed effects are due to physical, chemical, or biological effects. Baho et al. [87] observed that ecotoxicological effects were most frequently studied in bacteria, plants, and worms, while physical effects and reproductive impairment were the more widely reported. Also, community structure and functioning have been impaired by MPs, which is of great concern since soil microbial communities are essential for plant survival.

6.2 Chemical Effects

MPs can cause damage to biota, also due to their chemical characteristics. The backbone structure of MPs is the polymer that is usually considered chemically inert and therefore not hazardous. However, polymerization is often not complete, and a part of residual monomers can be found in the polymeric material, several of which are hazardous to human health and/or the environment [104]. Various other substances can be present in MPs, additives, plasticizers, oligomers, and solvents. Some plasticizer components were classified as endocrine disruptors (e.g., phthalates, bisphenol A) or toxic trace elements (e.g., chromium, cadmium, and lead) [105]. When larger plastic particles fragment into smaller pieces, an exponential increase in the surface/volume ratio increases the potential for leaching estrogenically active compounds [88].

In addition, trace elements and persistent organic pollutants (POPs), i.e., polycyclic aromatic hydrocarbons (PAHs), polychlorobiphenyls (PCBs), and the pesticide dichlorodiphenyltrichloroethane (DDT), could potentially be sorbed from water onto microplastic particles [106] and be transported [85, 107–109]. The subject is complex since the sorption of chemicals to MPs depends on several factors. Wang et al. [105] tried to clarify the interaction between chemicals and MPs, individuating the following factors: hydrophobicity of chemicals; density, chemical composition, and color of particles; size and weathering/aging of particles; and salinity, pH, and temperature of the water.

The so-called vector hypothesis is controversially discussed, and contradictory results can be found in the literature about the impact of microplastics through this specific pathway. MPs may alter the effect pollutants could have in marine waters, resulting in additional or reduced biological effects, whereas very little information is available for freshwater organisms. Oliveira et al. [110] exposed young freshwater gobies to pyrene sorbed on MPs and to pyrene alone. Fish exposed to pyrene sorbed on MPs showed a higher pyrene metabolite accumulation than fish exposed to pyrene alone. Besides, the combined exposure of MPs and pyrene caused inhibition of the isocitrate dehydrogenase enzyme, not observed following exposure to the two stressors separately. These results suggest a potential toxicological interaction of pollutants sorbed on MPs in the mechanisms leading to the enzymatic inhibition. However, the sorption on MPs can delay mortality, as shown by the study's data [110]. A possible cumulative effect was also observed in Guilhermino et al. [111], Grigorakis and Drouillard [112], and Rochman et al. [113] in freshwater species. Karami et al. [114] and Kleinteich et al. [115] suppose a possible reduction in the bioavailability of POPs, induced by MPs, with a consequent reduction in the POPs toxicity.

However, the application of different models for assessing the transport of POPs by MPs in marine water evidenced that if an effect exists, it is likely too small to be relevant from a risk assessment perspective [116, 117]. A preliminary study confirmed this hypothesis, observing no morphological effect on zebrafish exposed to benzo[*a*]pyrene sorbed on MPs [118]. The vector hypothesis is concerning because MPs are highly durable; hence, if they can carry high amounts of POPs, they could transfer pollutants from contaminated areas of water bodies to clean environments, where pollutants could be desorbed in the presence of different concentration organisms, not yet exposed to pollutants.

The vector hypothesis has been very little explored in the terrestrial environment, and the resulting picture is complex and variegate. Hodson et al. [119] demonstrated that microplastics could act as vectors of zinc in earthworms, although the associated risk is unlikely to be significant. Wang et al. [120] reported that microplastics neither act as a carrier of PAHs and PCBs nor contribute to bioaccumulation of these contaminants in earthworms. Hüffer et al. [121] demonstrated that the presence of PE microplastics in soil reduced the sorption of two pesticides by soil, thus suggesting that PE contamination may reduce the soil's natural retention capacity and increase the mobility of organic contaminants. There is a considerable lack of knowledge about the combined effect of microplastics and their additives on terrestrial biota [122].

6.3 Biological Effects

Microorganisms can colonize the surface of MPs [86], forming biofilms. It has been demonstrated in a freshwater study [123] that the taxonomic composition of these biofilms is distinct from the bacteria in the water column and suspended organic particulate, suggesting that MPs could select a specific suite of bacteria and become a vector of them. Scherer et al. [124] suggested that the colonization of MPs by microbes could distort the "taste" of MPs, making them more attractive or repulsing for aquatic biota. Biofilms can also change the plastic physical properties, i.e., density and hydrophobicity [74]. The increase in MP density could influence their transport in the water bodies, making the particles settle into the sediments. Less hydrophobic MPs could impact the sorption of organic chemicals, even if it has not yet been demonstrated [125].

The transport of microorganisms via MPs is of great concern because microorganisms could be transported at a great distance, giving rise to, for example, a possible introduction of pathogens in a clean environment. Potentially pathogenic species of microbes have already been found attached to microplastics [126].

As described in this section, the actual effects of microplastics in organisms have not yet been understood in depth. Many laboratory studies provided evidence for the potential effects of MPs on organisms, but many of them focus on single native polymers or narrow scale of particle size, shape, and duration of exposure [127]. Due to different exposure conditions and target species, some contradictory results are found in the literature. In future research, it will be essential to conduct laboratory studies, which mimic as much as possible the natural environment in terms of particle size, shape, polymer type, variety, and concentration, and to test the impact directly on the field.

7 Supportable Arrangements

Various technologies are being implemented for addressing the issue of MP occurrence in the environment. Current technologies deal with the problem mainly in two ways: (1) directly preventing plastic leakage into waterways and (2) collecting existing plastic pollution.

Schmaltz et al. [128] reviewed the existing technologies for reducing macro- and microplastics in the environment, evidencing how technologies aimed at preventing plastic leakage are fewer than those collecting it from marine and riverine environments. Some of the preventing technologies are dealing with avoiding microplastics from entering the water system through residential water. Among the preventing tools, some filters can be placed outside or inside the washing machine to capture microfibers generated during the washing process from synthetic clothes [128]. MPs can also be removed in the wastewater treatment plants (WWTPs) before entering the environment, even though WWTPs are not explicitly designed for MPs' removal. The primary treatment is sedimentation, which separates suspended solid particles from a liquid stream by gravitational settling; it is often associated with air flotation technology, consisting of microscopic air bubbles that enhance the natural tendency of contaminants to float on the surface before being collected. This stage can remove from 41 to 92% of MPs, depending on the characteristics of the wastewater flow. The secondary stage is capable of further removing MPs by 0.2–14% via biological treatment. The tertiary treatment further reduces the existence of MPs to 0.2–2% by various techniques (sand filters, membranes, etc.). Globally, MPs' removal in WWTPs worldwide varies from about 60% to 99.9%, depending upon the applied technology [129]. However, the efficiency of WWTPs is also a function of particle size, as small MPs and nanoplastics may not be effectively retained.

Advanced remediation technologies have been explored to degrade or remove the existing MPs from water. Even if plastic degradation has been studied for decades, strategies that focus on MPs are still limited. The existing technologies are mainly based on the following mechanisms: microbial degradation, advanced oxidation processes, thermal treatment, and other physical approaches such as adsorption and filtration [130].

Among the various mechanisms, biodegradation has the advantages of environmental friendliness, low cost, and low energy input. It has been demonstrated that bacteria and fungi can survive using plastics as carbon and nitrogen sources. The degradation of MPs by bacteria seems to be mostly very slow [131]. Fungi were reported to have a higher degradation activity toward LDPE than bacteria [132].

An example of an advanced oxidation process (AOP) is photocatalytic degradation. It can be induced by using a semiconductor photocatalyst, exposed to an appropriate light source, generating protons (H^+) and excited electrons (e^-). Protons produce various reactive oxygen species, which will attack MPs and cause polymeric chain rupture. This technique has been proved to be energy-efficient to degrade MPs in the aqueous environment. Other AOPs exploit the high redox potentials of peroxides (e. g., H_2O_2) to produce hydroxyl radicals for the degradation of MPs [130].

Thermal treatments are promising given the high calorific value of plastics, which can be converted from waste to fuel. Depending on the conversion method, plastics can be selectively converted either to bio-oil, bio-crude oil, synthesis gas, hydrogen, or aromatic char [133]. Examples of in-use MP collection techniques are the "Marine Microplastic Removal Tool" [134], which is a sand filter that can directly collect MPs, and the "Hoola One" [135], which can collect plastics by vacuuming sand and separating out macroplastics and microplastics by buoyancy. This technology reduces the occurrence of plastics to landfills and oceans, reduces carbon footprints, and has high conversion efficiency and cost-effectiveness.

References

1. Cera A, Cesarini G, Scalici M (2020) Microplastics in freshwater: what is the news from the world? Diversity 12(7):276. https://doi.org/10.3390/d12070276
2. Lu C, Di W, Peng L, Wenyou H, Li X, Yicheng S, Qiumei W, Kang T, Biao H, Seo Joon Y, Bong-Oh K, Jong Seong K (2021) Occurrence, distribution and affecting factors of microplastics in agricultural soils along the lower reaches of Yangtze River, China. Sci Total Environ 794:148694. https://doi.org/10.1016/j.scitotenv.2021.148694
3. Koutnik VS, Leonard J, Alkidim S et al (2021) Distribution of microplastics in soil and freshwater environments: global analysis and framework for transport modeling. Environ Pollut 274:116552
4. Gallagher A, Rees A, Rowe R, Stevens J, Wright P (2016) Microplastics in the Solent estuarine complex, UK: an initial assessment. Mar Pollut Bull 102(2):243–249. https://doi.org/10.1016/j.marpolbul.2015.04.002
5. Li J, Song Y, Cai Y (2020) Focus topics on microplastics in soil: analytical methods, occurrence, transport, and ecological risks. Environ Pollut 257:113570. https://doi.org/10.1016/j.envpol.2019.113570
6. Szymańska M, Obolewski K (2020) Microplastics as contaminants in freshwater environments: a multidisciplinary review. Ecohydrol Hydrobiol 20(3):333–345. https://doi.org/10.1016/j.ecohyd.2020.05.001
7. Yao L, Hui L, Yang Z, Chen X, Xiao A (2020) Freshwater microplastics pollution: detecting and visualizing emerging trends based on Citespace II. Chemosphere 245:125627. https://doi.org/10.1016/j.chemosphere.2019.125627. Epub 2019 Dec 11
8. Rasta M, Sattari M, Taleshi MS, Namin JI (2020) Identification and distribution of microplastics in the sediments and surface waters of Anzali Wetland in the Southwest Caspian Sea, Northern Iran. Mar Pollut Bull 160:111541. https://doi.org/10.1016/j.marpolbul.2020.111541
9. Lebreton LC, Van Der Zwet J, Damsteeg JW, Slat B, Andrady A, Reisser J (2017) River plastic emissions to the world's oceans. Nat Commun 8(1):15611. https://doi.org/10.1038/ncomms15611
10. Alimi OS, Farner Budarz J, Hernandez LM, Tufenkji N (2018) Microplastics and nanoplastics in aquatic environments: aggregation, deposition, and enhanced contaminant transport. Environ Sci Technol 52(4):1704–1724
11. Lechner A, Keckeis H, Lumesberger LF, Zens B, Krusch R, Tritthart M, Glas M, Schludermann E (2014) The Danube so colourful: a potpourri of plastic litter outnumbers

fish larvae in Europe's second largest river. Environ Pollut 188:177–181. https://doi.org/10.1016/j.envpol.2014.02.006
12. Stanton T, Johnson M, Nathanail P, MacNaughtan W, Gomes Rachel L (2020) Freshwater microplastic concentrations vary through both space and time. Environ Pollut 263(Part B):114481. https://doi.org/10.1016/j.envpol.2020.114481
13. Leslie HA, Brandsma SH, van Velzena MJM, Vethaakab AD (2017) Microplastics en route: field measurements in the Dutch river delta and Amsterdam canals, wastewater treatment plants, North Sea sediments and biota. Environ Int 101:133–142. https://doi.org/10.1016/j.envint.2017.01.018
14. Fu Z, Wang J (2019) Current practices and future perspectives of microplastic pollution in freshwater ecosystems in China. Sci Total Environ 691:697–712. https://doi.org/10.1016/j.scitotenv.2019.07.167. Epub 2019
15. Zhao S, Zhu L, Wang T, Li D (2014) Suspended microplastics in the surface water of the Yangtze estuary system, China: first observations on occurrence, distribution. Mar Pollut Bull 86:562–568
16. Fok L, Cheung P (2015) Hong Kong at the Pearl River estuary: a hotspot of microplastic pollution. Mar Pollut Bull 99:112–118
17. Zhang L, Xie Y, Zhong S, Liu J, Qin Y, Gao P (2021) Microplastics in freshwater and wild fishes from Lijiang River in Guangxi, Southwest China. Sci Total Environ 755:142428. https://doi.org/10.1016/j.scitotenv.2020.142428
18. Long Z, Pan Z, Wang W, Ren J, Yu X, Lin L, Lin H, Chen H, Jin X (2019) Microplastic abundance, characteristics, and removal in wastewater treatment plants in a coastal city of China. Water Res 155:255–265. https://doi.org/10.1016/j.watres.2019.02.028
19. Said L, Heard MJ (2020) Variation in the presence and abundance of anthropogenic microfibers in the Cumberland River in Nashville, TN, USA. Environ Sci Pollut Res 27: 10135–10139. https://doi.org/10.1007/s11356-020-08091-x
20. Leads RR, Weinstein JE (2019) Occurrence of tire wear particles and other microplastics within the tributaries of the Charleston Harbor Estuary, South Carolina, USA. Mar Pollut Bull 145:569–582. https://doi.org/10.1016/j.marpolbul.2019.06.061
21. Napper IE, Baroth A, Barrett AC, Bhola S, Chowdhury GW, Davies Bede FR, Duncan EM, Kumar S, Nelms SE, Niloy MN, Hasan NB, Maddalene T, Thompson RC, Koldewey H (2021) The abundance and characteristics of microplastics in surface water in the transboundary Ganges River. Environ Pollut 274:116348. https://doi.org/10.1016/j.envpol.2020.116348
22. Rech S, Macaya-Caquilpán V, Pantoja JF, Rivadeneira MM, Compodonico CK, Thiel M (2015) Sampling of riverine litter with citizen scientists — findings and recommendations. Environ Monit Assess 187:335. https://doi.org/10.1007/s10661-015-4473-y
23. Cózar A, Echevarría F, González-Gordillo JI, Irigoien X, Úbeda B, Hernández-León S, Palma AT, Navarro S, Garcia-de-Lomas J, Ruiz A, Fernandez-de-Puelles ML, Duarte CM (2014) Plastic debris in the open ocean. Proc Natl Acad Sci U S A 111(28):10239–10244. https://doi.org/10.1073/pnas.1314705111
24. Scherer C, Weber A, Lambert S, Wagner M (2018) Interactions of microplastics with freshwater biota. In: Freshwater microplastics. The handbook of environmental chemistry, vol 58. Springer, pp 153–180
25. Scherer C, Wolf R, Völker J et al (2020) Toxicity of microplastics and natural particles in the freshwater dipteran Chironomus riparius: same same but different? Sci Total Environ 711: 134604. https://doi.org/10.1016/j.scitotenv.2019.134604
26. Klein S, Worch E, Knepper TP (2015) Occurrence and spatial distribution of microplastics in river shore sediments of the Rhine-Main area in Germany. Environ Sci Technol 49(10):6070–6076
27. Mani T, Burkhardt-Holm P (2020) Seasonal microplastics variation in nival and pluvial stretches of the Rhine River–from the Swiss catchment towards the North Sea. Sci Total Environ 707:135579

28. Kiss T, Fórián S, Szatmári G, Sipos G (2021) Spatial distribution of microplastics in the fluvial sediments of a transboundary river – a case study of the Tisza River in Central Europe. Sci Total Environ 785:147306. https://doi.org/10.1016/j.scitotenv.2021.147306
29. Blair RM, Waldron S, Phoenix VR et al (2019) Microscopy and elemental analysis characterisation of microplastics in sediment of a freshwater urban river in Scotland, UK. Environ Sci Pollut Res 26:12491–12504. https://doi.org/10.1007/s11356-019-04678-1
30. Eriksen M, Mason S, Wilson S, Box C, Zellers A, Edwards W, Farley H, Amato S (2013) Microplastic pollution in the surface waters of the Laurentian Great Lakes. Mar Pollut Bull 77: 177–182. https://doi.org/10.1016/j.marpolbul.2013.10.007
31. Zbyszewski M, Corcoran PL (2011) Distribution and degradation of fresh water plastic particles along the beaches of Lake Huron, Canada. Water Air Soil Pollut 220:365–372. https://doi.org/10.1007/s11270-011-0760-6
32. Ballent A, Corcoran PL, Odile M, Helm PA, Longstaffe FJ (2016) Sources and sinks of microplastics in Canadian Lake Ontario nearshore, tributary and beach sediments. Mar Pollut Bull 110(1):383–395. https://doi.org/10.1016/j.marpolbul.2016.06.037
33. Lenaker PL, Corsi SR, Mason SA (2020) Spatial distribution of microplastics in surficial benthic sediment of Lake Michigan and Lake Erie. Environ Sci Technol 55(1):373–384. https://doi.org/10.1021/acs.est.0c06087
34. Baldwin AK, Spanjer AR, Rosen MR, Thom T (2020) Microplastics in Lake Mead National Recreation Area, USA: occurrence and biological uptake. PLoS One 15(5):e0228896
35. Fischer EK, Paglialonga L, Czech E, Tamminga M (2016) Microplastic pollution in lakes and lake shoreline sediments – a case study on Lake Bolsena and Lake Chiusi (central Italy). Environ Pollut 213:648–657. https://doi.org/10.1016/j.envpol.2016.03.012
36. Scopetani C, Chelazzi D, Cincinelli A, Esterhuizen-Londt M (2019) Assessment of microplastic pollution: occurrence and characterisation in Vesijärvi lake and Pikku Vesijärvi pond, Finland. Environ Monit Assess 191(11):652. https://doi.org/10.1016/j.etap.2019.03.007
37. Mehdinia A, Dehbandi R, Hamzehpour A, Rahnama R (2020) Identification of microplastics in the sediments of southern coasts of the Caspian Sea, north of Iran. Environ Pollut 258:113738. https://doi.org/10.1016/j.envpol.2019.113738
38. Zobkov M, Belkina N, Kovalevski V, Zobkova M, Efremova T, Galakhina N (2020) Microplastic abundance and accumulation behavior in Lake Onego sediments: a journey from the river mouth to pelagic waters of the large boreal lake. J Environ Chem Eng 8(5):104367. https://doi.org/10.1016/j.jece.2020.104367
39. Yuan W, Liu X, Wang W, Di M, Wang J (2019) Microplastic abundance, distribution and composition in water, sediments, and wild fish from Poyang Lake, China. Ecotoxicol Environ Saf 170:180–187. https://doi.org/10.1016/j.ecoenv.2018.11.126
40. Xiong X, Zhang K, Chen X, Shi H, Luo Z, Wu C (2018) Sources and distribution of microplastics in China's largest inland lake–Qinghai Lake. Environ Pollut 235:899–906. https://doi.org/10.1016/j.envpol.2017.12.081
41. Mao R, Song J, Yan P, Ouyang Z, Wu R, Liu S, Guo X (2021) Horizontal and vertical distribution of microplastics in the Wuliangsuhai Lake sediment, northern China. Sci Total Environ 754:142426. https://doi.org/10.1016/j.scitotenv.2020.142426
42. Qin Y, Wang Z, Li W, Chang X, Yang J, Yang F (2019) Microplastics in the sediment of Lake Ulansuhai of Yellow River Basin, China. Water Environ Res 92(6):829–839. https://doi.org/10.1002/wer.1275
43. Yin L, Wen X, Du C, Jiang J, Wu L, Zhang Y et al (2020) Comparison of the abundance of microplastics between rural and urban areas: a case study from East Dongting Lake. Chemosphere 244:125486. https://doi.org/10.1016/j.scitotenv.2016.03.053
44. Negrete Velasco ADJ, Rard L, Blois W, Lebrun D, Lebrun F, Pothe F, Stoll S (2020) Microplastic and fibre contamination in a remote mountain lake in Switzerland. Water 12 (9):2410. https://doi.org/10.3390/w12092410

45. Zhang K, Su J, Xiong X, Wu X, Wu C, Liu J (2016) Microplastic pollution of lakeshore sediments from remote lakes in Tibet plateau, China. Environ Pollut 219:450–455. https://doi.org/10.1016/j.envpol.2016.05.048
46. Free CM, Jensen OP, Mason SA, Eriksen M, Williamson NJ, Boldgiv B (2014) High-levels of microplastic pollution in a large, remote, mountain lake. Mar Pollut Bull 85(1):156–163. https://doi.org/10.1016/j.marpolbul.2014.06.001
47. Townsend KR, Lu HC, Sharley DJ, Pettigrove V (2019) Associations between microplastic pollution and land use in urban wetland sediments. Environ Sci Pollut Res 26:22551–22561. https://doi.org/10.1007/s11356-019-04885-w
48. Triebskorn R, Braunbeck T, Grummt T, Hanslik L, Huppertsberg S, Jekel M, Knepper TP, Krais S, Müller YK, Pittroff M, Ruhl AS, Schmieg H, Schür C, Strobel C, Wagner M, Zumbülte N, Köhler H (2019) Relevance of nano- and microplastics for freshwater ecosystems: a critical review. TrAC Trends Anal Chem 110:375–392. https://doi.org/10.1016/j.trac.2018.11.023
49. Lin L, Pan X, Zhang S, Li D, Zhai W, Wang Z et al (2021) Distribution and source of microplastics in China's second largest reservoir-Danjiangkou Reservoir. J Environ Sci 102:74–84. https://doi.org/10.1016/j.jes.2020.09.018
50. Liu K, Zhang F, Song Z, Zong C, Wei N, Li D (2019) A novel method enabling the accurate quantification of microplastics in the water column of deep ocean. Mar Pollut Bull 146:462–465. https://doi.org/10.1016/j.marpolbul.2019.07.008
51. Olesen KB, Stephansen DA, van Alst N, Vollertsen J (2019) Microplastics in a stormwater pond. Water 11:1466. https://doi.org/10.3390/w11071466
52. Ziajahromi S, Neale PA, Telles Silveira I, Chua A, Leusch FDL (2021) An audit of microplastic abundance throughout three Australian wastewater treatment plants. Chemosphere 263:128294. https://doi.org/10.1016/j.chemosphere.2020.128294
53. Knight LJ, Parker-Jurd FNF, Al-Sid-Cheikh M, Thompson RC (2020) Tyre wear particles: an abundant yet widely unreported microplastic? Environ Sci Pollut Res 27:18345–18354. https://doi.org/10.1007/s11356-020-08187-4
54. Guo J-J, Huang X-P, Xiang L, Wang Y-Z, Li Y-W, Li H, Cai Q-Y, Mo C-H, Wong M-H (2020) Source, migration and toxicology of microplastics in soil. Environ Int 137:105263. https://doi.org/10.1016/j.envint.2019.105263
55. Rafique A, Irfan M, Mumtaz M et al (2020) Spatial distribution of microplastics in soil with context to human activities: a case study from the urban center. Environ Monit Assess 192:671. https://doi.org/10.1007/s10661-020-08641-3
56. Fuller S, Gautam AA (2020) Procedure for measuring microplastics using pressurized fluid extraction. Environ Sci Technol 11(2016):5774–5780. https://doi.org/10.1021/acs.est.6b00816
57. Bläsing M, Amelung W (2018) Plastics in soil: analytical methods and possible sources. Sci Total Environ 612(2018):422–435. https://doi.org/10.1016/j.scitotenv.2017.08.086
58. Zhang GS, Liu YF (2018) The distribution of microplastics in soil aggregate fractions in southwestern China. Sci Total Environ 642:12–20. https://doi.org/10.1016/j.scitotenv.2018.06.004. Epub 2018 Jun 9
59. Kumar M, Xiong X, He M, Tsang Daniel CW, Gupta J, Khan E, Harrad S, Hou D, Ok YS, Bolan NS (2020) Microplastics as pollutants in agricultural soils. Environ Pollut 265(Part A):114980. https://doi.org/10.1016/j.envpol.2020.114980
60. Talvitie J, Mikola A, Setala O, Heinonen M, Koistinen A (2017) How well is microlitter purified from wastewater? A detailed study on the stepwise removal of microlitter in a tertiary level wastewater treatment plant. Water Res 109:164–172. https://doi.org/10.1016/j.watres.2016.11.046
61. van den Berg P, Huerta-Lwanga E, Corradini F, Geissen V (2020) Sewage sludge application as a vehicle for microplastics in eastern Spanish agricultural soils. Environ Pollut 261:114198. https://doi.org/10.1016/j.envpol.2020.114198

62. Zubris KAV, Richards BK (2005) Synthetic fibers as an indicator of land application of sludge. Environ Pollut 138(2):201–211. https://doi.org/10.1016/j.envpol.2005.04.013
63. Corradini F, Meza P, Eguiluz R, Casado F, Huerta-Lwanga E, Geissen V (2019) Evidence of microplastic accumulation in agricultural soils from sewage sludge disposal. Sci Total Environ 671:411–420. https://doi.org/10.1016/j.scitotenv.2019.03.368
64. Gui J, Sun Y, Wang J, Chen X, Zhang S, Wu D (2021) Microplastics in composting of rural domestic waste: abundance, characteristics, and release from the surface of macroplastics. Environ Pollut 274:116553. https://doi.org/10.1016/j.envpol.2021.116553
65. Weithmann N, Möller JN, Löder MGJ, Piehl S, Laforsch C, Freitag R (2018) Organic fertilizer as a vehicle for the entry of microplastic into the environment. Sci Adv 4:8060. https://doi.org/10.1126/sciadv.aap8060
66. Kilponen J (2016) Microplastics and harmful substances in urban runoffs and landfill leachates: possible emission sources to marine environment, Finnish Environment Institute. Lahti Univ Appl Sci
67. Su Y, Zhang Zhongjian W, Dong ZL, Shi H, Xie B (2019) Occurrence of microplastics in landfill systems and their fate with landfill age. Water Res 164:114968. https://doi.org/10.1016/j.watres.2019.114968
68. He P, Chen L, Shao L, Zhang H, Lü F (2019) Municipal solid waste (MSW) landfill: a source of microplastics? – Evidence of microplastics in landfill leachate. Water Res 159:38–45. https://doi.org/10.1016/j.watres.2019.04.060
69. Lares M, Ncibi MC, Sillanpää M, Sillanpää M (2018) Occurrence, identification and removal of microplastic particles and fibers in conventional activated sludge process and advanced MBR technology. Water Res 133:236–246. https://doi.org/10.1016/j.watres.2018.01.049
70. Tibbetts J, Krause S, Lynch I, Smith GHS (2018) Abundance, distribution, and drivers of microplastic contamination in urban river environments. Water 1597(10):1597. https://doi.org/10.3390/W10111597
71. Koelmans AA, Quik JTK, Velzeboer I (2015) Lake retention of manufactured nanoparticles. Environ Pollut 196:171–175. https://doi.org/10.1016/j.envpol.2014.09.025
72. Kooi M, Besseling E, Kroeze C et al (2018) Modeling the fate and transport of plastic debris in freshwaters: review and guidance. In: Handbook of environmental chemistry. Springer, pp 125–152
73. Qu HJ, Kroeze C (2011) Nutrient export by rivers to the coastal waters of China: management strategies and future trends. Reg Environ Chang 121(12):153–167. https://doi.org/10.1007/S10113-011-0248-3
74. Carr SA, Liu J, Tesoro AG (2016) Transport and fate of microplastic particles in wastewater treatment plants. Water Res 91:174–182. https://doi.org/10.1016/j.watres.2016.01.002
75. Rochman CM (2018) Microplastics research — from sink to source in freshwater systems. Science 360:28–29
76. Sanchez C (2020) Fungal potential for the degradation of petroleum-based polymers: an overview of macro- and microplastics biodegradation. Biotechnol Adv 40:107501. https://doi.org/10.1016/j.biotechadv.2019.107501
77. Rillig MC, Ingraffia R, De Souza Machado AA (2017) Microplastic incorporation into soil in agroecosystems. Front Plant Sci 8. https://doi.org/10.3389/fpls.2017.01805
78. Huerta Lwanga E, Gertsen H, Gooren H et al (2016) Microplastics in the terrestrial ecosystem: implications for Lumbricus terrestris (Oligochaeta, Lumbricidae). Environ Sci Technol 50: 2685–2691. https://doi.org/10.1021/acs.est.5b05478
79. Rillig MC, Ziersch L, Hempel S (2017) Microplastic transport in soil by earthworms. Sci Rep 7. https://doi.org/10.1038/s41598-017-01594-7
80. Horton AA, Walton A, Spurgeon DJ et al (2017) Microplastics in freshwater and terrestrial environments: evaluating the current understanding to identify the knowledge gaps and future research priorities. Sci Total Environ 586:127–141
81. Wang C, Zhao J, Xing B (2021) Environmental source, fate, and toxicity of microplastics. J Hazard Mater 407:124357

82. O'Connor D, Pan S, Shen Z et al (2019) Microplastics undergo accelerated vertical migration in sand soil due to small size and wet-dry cycles. Environ Pollut 249:527–534. https://doi.org/10.1016/j.envpol.2019.03.092
83. Liu G, Jiang R, You J et al (2020) Microplastic impacts on microalgae growth: effects of size and humic acid. Environ Sci Technol 54:1782–1789. https://doi.org/10.1021/acs.est.9b06187
84. Sjollema SB, Redondo-Hasselerharm P, Leslie HA et al (2016) Do plastic particles affect microalgal photosynthesis and growth? Aquat Toxicol 170:259–261. https://doi.org/10.1016/j.aquatox.2015.12.002
85. Brennecke D, Duarte B, Paiva F et al (2016) Microplastics as vector for heavy metal contamination from the marine environment. Estuar Coast Shelf Sci 178:189–195. https://doi.org/10.1016/j.ecss.2015.12.003
86. Harrison JP, Schratzberger M, Sapp M, Osborn AM (2014) Rapid bacterial colonization of low-density polyethylene microplastics in coastal sediment microcosms. BMC Microbiol 14: 1–15. https://doi.org/10.1186/s12866-014-0232-4
87. Baho DL, Bundschuh M, Futter MN (2021) Microplastics in terrestrial ecosystems: moving beyond the state of the art to minimize the risk of ecological surprise. Glob Chang Biol:1–18. https://doi.org/10.1111/gcb.15724
88. de Souza Machado AA, Kloas W, Zarfl C et al (2018) Microplastics as an emerging threat to terrestrial ecosystems. Glob Chang Biol 24:1405–1416. https://doi.org/10.1111/gcb.14020
89. Eerkes-Medrano D, Thompson RC, Aldridge DC (2015) Microplastics in freshwater systems: a review of the emerging threats, identification of knowledge gaps and prioritisation of research needs. Water Res 75:63–82
90. Zhao S, Zhu L, Li D (2016) Microscopic anthropogenic litter in terrestrial birds from Shanghai, China: not only plastics but also natural fibers. Sci Total Environ 550:1110–1115. https://doi.org/10.1016/j.scitotenv.2016.01.112
91. De Souza Machado AA, Lau CW, Kloas W et al (2019) Microplastics can change soil properties and affect plant performance. Environ Sci Technol 53:6044–6052. https://doi.org/10.1021/acs.est.9b01339
92. Jin Y, Lu L, Tu W et al (2019) Impacts of polystyrene microplastic on the gut barrier, microbiota and metabolism of mice. Sci Total Environ 649:308–317. https://doi.org/10.1016/j.scitotenv.2018.08.353
93. Qi Y, Yang X, Pelaez AM et al (2018) Macro- and micro-plastics in soil-plant system: effects of plastic mulch film residues on wheat (Triticum aestivum) growth. Sci Total Environ 645: 1048–1056. https://doi.org/10.1016/j.scitotenv.2018.07.229
94. Laist DW (1996) Marine debris entanglement and ghost fishing: a cryptic and significant type of bycatch? In: Solving bycatch: consideration for today and tomorrow, pp 33–39
95. Ryan PG (2018) Entanglement of birds in plastics and other synthetic materials. Mar Pollut Bull 135:159–164. https://doi.org/10.1016/j.marpolbul.2018.06.057
96. Eriksson C, Burton H (2003) Origins and biological accumulation of small plastic particles in fur seals from Macquarie Island. AMBIO A J Hum Environ 32:380. https://doi.org/10.1639/0044-7447(2003)032[0380:oabaos]2.0.co;2
97. Huerta Lwanga E, Mendoza Vega J, Ku Quej V et al (2017) Field evidence for transfer of plastic debris along a terrestrial food chain. Sci Rep 7. https://doi.org/10.1038/s41598-017-14588-2
98. Au SY, Bruce TF, Bridges WC, Klaine SJ (2015) Responses of Hyalella azteca to acute and chronic microplastic exposures. Environ Toxicol Chem 34(11):2564–2572. https://doi.org/10.1002/etc.3093
99. Ogonowski M, Schür C, Jarsén Å, Gorokhova E (2016) The effects of natural and anthropogenic microparticles on individual fitness in daphnia magna. PLoS One 11:155063. https://doi.org/10.1371/journal.pone.0155063
100. Rehse S, Kloas W, Zarfl C (2016) Short-term exposure with high concentrations of pristine microplastic particles leads to immobilisation of Daphnia magna. Chemosphere 153:91–99. https://doi.org/10.1016/j.chemosphere.2016.02.133

101. Lu Y, Zhang Y, Deng Y et al (2016) Uptake and accumulation of polystyrene microplastics in zebrafish (Danio rerio) and toxic effects in liver. Environ Sci Technol 50:4054–4060. https://doi.org/10.1021/acs.est.6b00183
102. Imhof HK, Laforsch C (2016) Hazardous or not – are adult and juvenile individuals of Potamopyrgus antipodarum affected by non-buoyant microplastic particles? Environ Pollut 218:383–391. https://doi.org/10.1016/j.envpol.2016.07.017
103. Weber A, Scherer C, Brennholt N et al (2018) PET microplastics do not negatively affect the survival, development, metabolism and feeding activity of the freshwater invertebrate Gammarus pulex. Environ Pollut 234:181–189. https://doi.org/10.1016/j.envpol.2017.11.014
104. Lithner D, Larsson A, Dave G (2011) Environmental and health hazard ranking and assessment of plastic polymers based on chemical composition. Sci Total Environ 409:3309–3324. https://doi.org/10.1016/j.scitotenv.2011.04.038
105. Wang F, Wong CS, Chen D et al (2018) Interaction of toxic chemicals with microplastics: a critical review. Water Res 139:208–219
106. Bakir A, Rowland SJ, Thompson RC (2012) Competitive sorption of persistent organic pollutants onto microplastics in the marine environment. Mar Pollut Bull 64:2782–2789. https://doi.org/10.1016/j.marpolbul.2012.09.010
107. Bakir A, Rowland SJ, Thompson RC (2014) Transport of persistent organic pollutants by microplastics in estuarine conditions. Estuar Coast Shelf Sci 140:14–21. https://doi.org/10.1016/j.ecss.2014.01.004
108. Hirai H, Takada H, Ogata Y et al (2011) Organic micropollutants in marine plastics debris from the open ocean and remote and urban beaches. Mar Pollut Bull 62:1683–1692. https://doi.org/10.1016/j.marpolbul.2011.06.004
109. Holmes LA, Turner A, Thompson RC (2014) Interactions between trace metals and plastic production pellets under estuarine conditions. Mar Chem 167:25–32. https://doi.org/10.1016/j.marchem.2014.06.001
110. Oliveira M, Ribeiro A, Hylland K, Guilhermino L (2013) Single and combined effects of microplastics and pyrene on juveniles (0+ group) of the common goby Pomatoschistus microps (Teleostei, Gobiidae). Ecol Indic 34:641–647. https://doi.org/10.1016/j.ecolind.2013.06.019
111. Guilhermino L, Vieira LR, Ribeiro D et al (2018) Uptake and effects of the antimicrobial florfenicol, microplastics and their mixtures on freshwater exotic invasive bivalve Corbicula fluminea. Sci Total Environ 622–623:1131–1142. https://doi.org/10.1016/j.scitotenv.2017.12.020
112. Grigorakis S, Drouillard KG (2018) Effect of microplastic amendment to food on diet assimilation efficiencies of PCBs by fish. Environ Sci Technol 52:10796–10802. https://doi.org/10.1021/acs.est.8b02497
113. Rochman CM, Hoh E, Kurobe T, Teh SJ (2013) Ingested plastic transfers hazardous chemicals to fish and induces hepatic stress. Sci Rep 3:1–7. https://doi.org/10.1038/srep03263
114. Karami A, Romano N, Galloway T, Hamzah H (2016) Virgin microplastics cause toxicity and modulate the impacts of phenanthrene on biomarker responses in African catfish (Clarias gariepinus). Environ Res 151:58–70. https://doi.org/10.1016/j.envres.2016.07.024
115. Kleinteich J, Seidensticker S, Marggrander N, Zarf C (2018) Microplastics reduce short-term effects of environmental contaminants. Part II: Polyethylene particles decrease the effect of polycyclic aromatic hydrocarbons on microorganisms. Int J Environ Res Public Health 15: 287. https://doi.org/10.3390/ijerph15020287
116. Gouin T, Roche N, Lohmann R, Hodges G (2011) A thermodynamic approach for assessing the environmental exposure of chemicals absorbed to microplastic. Environ Sci Technol 45: 1466–1472. https://doi.org/10.1021/es1032025
117. Koelmans AA, Besseling E, Wegner A, Foekema EM (2013) Plastic as a carrier of POPs to aquatic organisms: a model analysis. Environ Sci Technol 47:7812–7820. https://doi.org/10.1021/es401169n

118. Batel A, Borchert F, Reinwald H et al (2018) Microplastic accumulation patterns and transfer of benzo[a]pyrene to adult zebrafish (Danio rerio) gills and zebrafish embryos. Environ Pollut 235:918–930. https://doi.org/10.1016/j.envpol.2018.01.028
119. Hodson ME, Duffus-Hodson CA, Clark A et al (2017) Plastic bag derived-microplastics as a vector for metal exposure in terrestrial invertebrates. Environ Sci Technol 51:4714–4721. https://doi.org/10.1021/ACS.EST.7B00635
120. Wang J, Coffin S, Sun C et al (2019) Negligible effects of microplastics on animal fitness and HOC bioaccumulation in earthworm Eisenia fetida in soil. Environ Pollut 249:776–784. https://doi.org/10.1016/J.ENVPOL.2019.03.102
121. Hüffer T, Metzelder F, Sigmund G et al (2019) Polyethylene microplastics influence the transport of organic contaminants in soil. Sci Total Environ 657:242–247. https://doi.org/10.1016/J.SCITOTENV.2018.12.047
122. Campanale C, Massarelli C, Savino I et al (2020) A detailed review study on potential effects of microplastics and additives of concern on human health. Int J Environ Res Public Health 17. https://doi.org/10.3390/IJERPH17041212
123. McCormick A, Hoellein TJ, Mason SA et al (2014) Microplastic is an abundant and distinct microbial habitat in an urban river. Environ Sci Technol 48:11863–11871. https://doi.org/10.1021/es503610r
124. Scherer C, Weber A, Stock F, Vurusic S, Egerci H, Kochleus C, Arendt N, Foeldi C, Dierkes G, Wagner M, Brennholt N, Reifferscheid G (2020) Comparative assessment of microplastics in water and sediment of a large European river. Sci Total Environ 738: 139866. https://doi.org/10.1016/j.scitotenv.2020.139866
125. Li T-C, Yuan C-S, Hung C-H et al (2016) Chemical characteristics of marine fine aerosols over sea and at offshore islands during three cruise sampling campaigns in the Taiwan Strait – Sea Salts and Anthropogenic Particles. Atmos Chem Phys Discuss 1–27. https://doi.org/10.5194/acp-2016-384
126. Jiang P, Zhao S, Zhu L, Li D (2018) Microplastic-associated bacterial assemblages in the intertidal zone of the Yangtze Estuary. Sci Total Environ 624:48–54. https://doi.org/10.1016/j.scitotenv.2017.12.105
127. Brennholt N, Heß M, Reifferscheid G (2018) Freshwater microplastics: challenges for regulation and management. In: Freshwater microplastics. The handbook of environmental chemistry, vol 58. Springer, pp 239–272
128. Schmaltz E, Melvin EC, Diana Z et al (2020) Plastic pollution solutions: emerging technologies to prevent and collect marine plastic pollution. Environ Int 144:106067
129. Ngo PL, Pramanik BK, Shah K, Roychand R (2019) Pathway, classification and removal efficiency of microplastics in wastewater treatment plants. Environ Pollut 255:113326. https://doi.org/10.1016/j.envpol.2019.113326
130. Hu K, Tian W, Yang Y et al (2021) Microplastics remediation in aqueous systems: strategies and technologies. Water Res 198:117144. https://doi.org/10.1016/j.watres.2021.117144
131. Roager L, Sonnenschein EC (2019) Bacterial candidates for colonization and degradation of marine plastic debris. Environ Sci Technol 53:11636–11643. https://doi.org/10.1021/acs.est.9b02212
132. Muhonja CN, Makonde H, Magoma G, Imbuga M (2018) Biodegradability of polyethylene by bacteria and fungi from Dandora dumpsite Nairobi-Kenya. PLoS One 13:e0198446
133. Nanda S, Berruti F (2021) Thermochemical conversion of plastic waste to fuels: a review. Environ Chem Lett 19:123–148. https://doi.org/10.1007/s10311-020-01094-7
134. Ward M (2013) Marine Microplastics Removal Tool Patent US8944253B2 US
135. Hoola One—We hand the beaches back to nature (n.d.) Hoola One. https://hoolaone.com. Accessed 18 Feb 2020

Agricultural Plastic Mulching as a Source of Microplastics in the Terrestrial Conditions

Mónica Calero, Mario J. Muñoz, Rafael R. Solís, Emilio J. Lozano, Verónica Godoy, and Mª. Ángeles Martín-Lara

1 The Role of Plastics in Agriculture: Benefits and Uses

Plasticulture is a well-known term coined to describe the intensive use of plastics in agriculture. The implementation of plastics for agricultural purposes is widespread and represents a growing market since their use is vital to harvest the production, raise the quality, or minimize the demand for some resources such as water or pesticides. Plastics help to control diverse livestock variables such as the temperature, light, moisture, irrigation, growth of weeds, and the apparition of pests, among others [1, 2]. Furthermore, watering can be considerably reduced due to lesser evaporation and a crop protected with plastics contributes to saving in fertilizers and pesticides. However, the use of plastics, although covering mostly plastic films in greenhouses and mulching, includes wider purposes such as nets, pipes, and small tunnels. In Europe, a demand for 322,500 tons of plastic was estimated in 2019 for different applications as shown in Fig. 1. From all of them, the larger amount of plastics was for greenhouses, mulching, and small tunnels: in other words plastic film, which accounted for 80% of the whole agriculture demand [3].

The use of plastics was firstly developed to increase and better control the soil temperature in the 1950s. Plastic mulching is the best-optimized technology for crop control [2] and is widely spread around the world due to the optimal microclimate created. The consumption of plastic film as mulches is steadily increasing in some countries such as Germany, Spain, and China [4]. The controlled temperature, especially at night, impacts positively other livestock variables, such as moisture [4], the nutrient uptake [5, 6], or the beneficial microorganisms colony in the solid

M. Calero (✉) · M. J. Muñoz · R. R. Solís · E. J. Lozano · V. Godoy · M. Á. Martín-Lara
Department of Chemical Engineering, Faculty of Sciences, University of Granada, Granada, Spain
e-mail: mcaleroh@ugr.es; mariomunoz@ugr.es; rafarsolis@ugr.es; emiliolp@ugr.es; vgcalero@ugr.es; marianml@ugr.es

A. Khan et al. (eds.), *Microplastic sources, fate and solution*,
https://doi.org/10.1007/978-981-99-0695-6_3

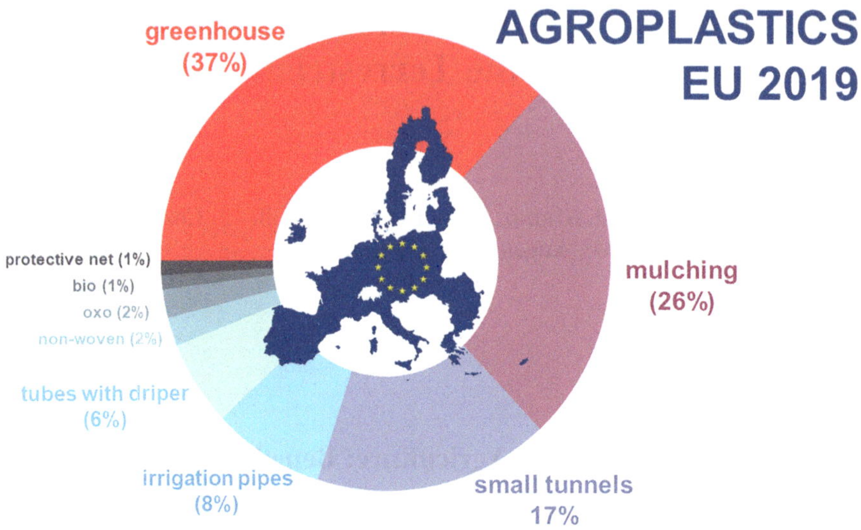

Fig. 1 Destination of the plastics used in agriculture in the European Union during 2019. *Adapted from Agriculture Plastic Environment Europe* [3]

[7], resulting in greater growth of the cultivated plants [8]. Moreover, the growth of pathogens is minimized and more efficiently controlled. Another attractive consequence of plastic covering is the reduced fertilizing requirements [5], which highly impacts the economy of the production. The radiation absorption and transmittance of plastic is a key design variable for the control of the generated microclimate [9]. Thus, black plastics can absorb radiation and transmit it as heat. However, if the plastic is transparent, the radiation passes through and may transmit to the soil within 85–95% of the inlet radiation. In cold regions, transparent plastics are preferred [10], whereas black plastics are popular in those regions in which temperatures can reach high values [11]. Sometimes, the objective might be the elimination of soil-borne pathogens by the maximum uptake of solar radiation [8]. In this case, known as solarization [12], transparent plastic can be used in hotter regions and high temperatures can be reached for this purpose, leading to controlled sterilization.

Plastic mulching can act as a barrier and control the watering needs [13]. This aspect is fundamental in those regions with high water stress. The plastic film acts as a barrier that impedes water loss by evapotranspiration which is translated into more water availability for the plant [14]. Mulching conserves soil moisture and controls soil erosion [15]. The percentage of water-saving uptake depends on the color of the plastic, i.e., estimated as ca. 26% for transparent white plastics and 33% for black plastics [14].

Another important role of plastic mulches is weed control. This strategy does not consume chemical herbicides and supports the preferred demand for pesticide-free food [16]. Plastic mulches provide a range of weed control levels, depending on the amount of light transmission through the mulch [17]. Black mulches control more

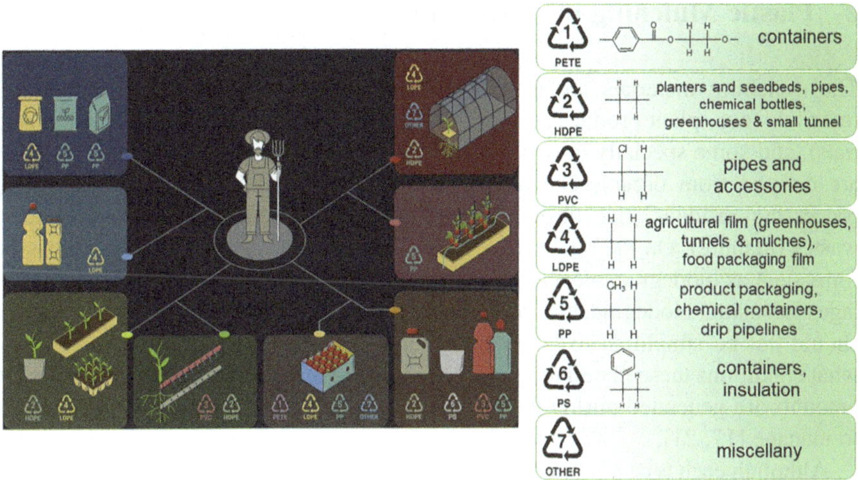

Fig. 2 Classification of plastics used in agriculture by their use and nature. Adapted from [19] with permission of Elsevier© 2021

efficiently the growth of weed; nevertheless, they provide less soil warming. Furthermore, mulches protect the crop from insects, pests, parasitic nematodes, and viruses that can trigger diverse diseases [2, 17].

Plastics are largely obtained in the petrochemical-derived industries, although the production from bio-based materials is taking precedence due to the environmentally oriented new policies in the developed countries. Plastics can be manufactured from a polymer or as a mixture. According to Plastics Europe, during the year 2019, the agriculture sector demanded 3.4% of the global plastic market [18]. Although a wide range of plastics is used in agriculture, the most popular are the poly-olefins such as polyethylene (PE) and polypropylene (PP) [1]. PP is harder and more heat resistant than PE. For that reason, PE can be used as low-density polyethylene (LDPE) or high-density polyethylene (HDPE) to increase the hardness and thermal resistance. Ethylene-vinyl acetate copolymer (EVA) is another polyolefin with relatively low consumption. Other examples of plastics with much lower demand are polyvinyl chloride (PVC), polyethylene terephthalate (PET), polycarbonate (PC), poly-methyl-methacrylate (PMMA), expanded polystyrene (EPS), etc. [19]. A vast list of plastic tools is essential in current agriculture. As summarized in Fig. 2, there are diverse applications of plastic as a source, for example, bags for fertilizers, seed or feed, containers for agrochemicals and products, plastic pipes for drainage and irrigation, nets, mesh, strings, ropes, trays, and more importantly, plastic films for covering greenhouses, small tunnels, mulching, and packaging [19]. The consumption of plastic film outshines the rest by far. Usually, polyethylene is used under the low-density version (LDPE) for making plastic films followed by polypropylene for a woven plastic film if a harder and more heat-resistant film is required.

2 Plastic Mulching in Agriculture

The mulching materials are usually classified into three main groups: organic materials (e.g., plant products, animal wastes), inorganic materials (synthetic materials), and some specially designed materials [20]. The organic mulching materials are derived from organic substances such as wood industrial wastes (sawdust), processing residues (rice husks), agricultural wastes (straw, stalks), and animal wastes (manure). On the other hand, specially designed materials are new types of films such as biodegradable and photodegradable plastic films [17, 20]. The advantages of using biodegradable polymers are well defined; however, the analysis should not be simplified to laboratory results, since it is yet to be defined under what conditions these materials are fully eliminated or if, on the contrary, significant amounts are generated during their degradation, micro- or nano-fragments difficult to manage [17, 21].

Although each type of mulching material has particular characteristics and is used for specific applications, the inorganic group of mulching materials (non-self-degradable plastic mulching) is widely used in modern agriculture [17, 20, 21]. The inorganic plastic mulching can be classified according to its chemical, optical, and morphologic properties.

Clear/Transparent Mulch This kind of mulch has no additives (e.g., pigments) to modify its optical properties. It is used to improve the soil temperature which in certain conditions drives a faster root development of the plants. The transparent mulch prevents water from evaporating rapidly from the surface of the soil as well ([22]; [23]; [24]).

Black Mulch (Total Opaqueness) This type of mulch is one of the most commonly used types of mulching, mainly due to its optical properties, defined by the black color. It prevents the natural sunlight from reaching the soil and allows better control of the weeds, avoiding water evaporation from the ground. The black structure absorbs a large amount of energy from sunlight, increasing its temperature during the day and then transferring it to the soil by direct contact. That kind of interaction provides a relatively good homogeneity of the temperature which allows a regular development of the root areas as well as plant growth. However, it has some limitations related to the temperate gradient, since it is very likely to cause burns to the organs of some delicate plants that may come into contact with the hot film, according to some reports ([20]; [17]; [25]).

Multicolor Mulch (Silver/Black or White/Black) The multicolor mulch is designed mainly to guarantee that weeds are kept under control. The black surface is placed downward to block out any light reaching the soil. The silver surface, with good reflection properties, is placed upward. In this way, the soil absorbs less heat due to reflection, but it compensates with the extra light radiation, which increases the process of photosynthesis and ensures faster plant growth. The reflection response of these films also acts as an insect repellent which helps in reducing the number of pesticides used ([26]; [27]; [28]).

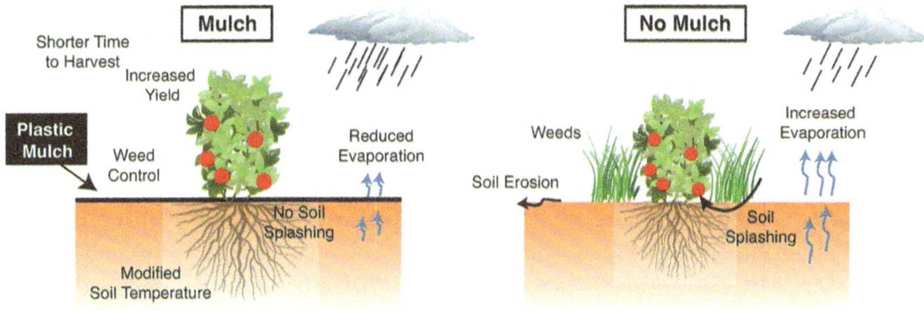

Fig. 3 Schematic representation of agricultural cropping system using and without plastic mulch film [21]

Photo-selective Mulches (Other Colors) This type of mulch is used to naturally control weeds through selective blocking of light radiation and at the same time to raise the temperature in the soil since the plastic's translucent color allows light and heat to directly reach the soil. These plastics offer the best combination of clear mulch and black mulch when soil heating is needed but farmers also want to refrain from the use of chemicals for weed control ([20]; [22]; [24]).

As it was explained before, different strategies in terms of optical response and structure can be used for a specific crop. The general advantages of mulching can be schematically described as shown in Fig. 3 [21].

The profits are (1) reduction of soil moisture loss, soil compaction, soil erosion, the impact of water droplets hitting the soil surface, weed growth, and competition for water and nutrients from the surrounding fields, (2) encouragement of earthworm movement into the soil that helps to improve soil structure and nutrient cycling, (3) keeping the underlying root zone warmer in winter and cooler in summer to reduce root damage, (4) providing a barrier for trees against lawnmowers and weed whips, (5) helping top growth and root development of plants, (6) improving the appearance of newly planted trees, (7) lowering soil pH to enhance crop root's ability to take up nutrients, and (8) increasing long-term nutrient availability in the soil [20].

In any case, the process occurring under solar irradiation and plastic mulches under irradiation is complex and provides in many cases contradictory reports as is previously analyzed by Zhang et al. [29]. These authors describe how the transparent plastic film mulch use results in higher soil temperature than the black mulches in the case of taro grown in southeastern Nigeria and watermelons grown in Croatia and south-western Mexico. Meanwhile, the black plastic film mulch leads to a higher soil temperature than the transparent plastic film mulch for potatoes grown in Spain and watermelon grown in Croatia. Some crop yields with black mulch are higher than with transparent mulch such as taro and potato. However, other crop yields with black plastic film mulch are lower than with transparent plastic film mulch such as watermelon and cucumber. Black plastic film mulch is better than the transparent/

clear plastic film mulch for high yield in some years, while the transparent plastic film mulch is better in other years at the same place. The yield with transparent plastic film mulch is even lower than without mulch [29].

In any case, under optimized operating conditions, considering the short-term effects on soil conditions (control on temperature and moisture), the plastic mulching provides higher product yield and quality, and consequently, higher economic value for farmers. On the other hand, the long-term effects on soil quality, farm profitability, and environmental pollution and ecosystem services need to be considered on public awareness and farmers' perceptions [17, 20]. Such will be analyzed in the following sections.

3 Plastic Mulch Waste

Residues from plastics and plastic elements used in agriculture for crop protection can harm the environment after use if not properly managed, as over time and with exposure to the sun, plastic decomposes into smaller particles that are absorbed by the soil, transported by water or wind, and also ingested by vertebrate and invertebrate animals.

Mulch film is creating a major environmental problem that is of concern to the sector and public administrations. Plastic elimination is both laborious and costly because it is a material composed mainly of low-density polyethylene (LDPE) and takes a long time to degrade. It is also often heavily contaminated with soil and plant residues, making it difficult to recycle [30].

The average soilage content across all products is estimated to have a coefficient of 1.55, resulting in 713K tonnes of new agri-plastics becoming 1175K tonnes of used agri-plastic waste (Table 1). One-third of this additional volume comes from

Table 1 The European market for agricultural plastics: new vs used [30]

Products	Tonnes new	%	Tonnes used	%
Films	547,000	76	927,500	80
Flexible irrigation pipes	40,000	5	48,000	4
Nets	54,500	8	80,500	7
Twines	80,000	11	104,000	9
Total	**721,500**	**100**	**1,160,000**	**100**
Polymers				
LDPE	587,000	81	975,500	84
HDPE	54,500	8	80,500	7
PP	80,000	11	104,000	9
Total	**721,500**	**100**	**1,160,000**	**100**
Without oxo	713,500			

Note: Tonnes used = Estimate of total plastic waste, including soilage content

products dedicated to animal production and two-thirds for vegetable production [30].

The terms macroplastic and microplastic are generally related to marine pollution, and numerous studies are describing the fate and effects of these materials in aquatic environments [31], but the soil is also one of the places with a large number of microplastics. Therefore, recent studies have evaluated the potential for soil contamination due to agricultural plastic mulching as a source of microplastics in the terrestrial environment [32].

According to Agriculture Plastics Environment (APE) Europe, the mulch film market in Europe amounts to about 83,000 tonnes, where more than 30% remains in the field and the soil, generating about 15,000 tonnes of microplastics each year.

The agricultural system plays an important role, as it could substantially affect the accumulation and distribution of agricultural plastic waste. The continuous use of plastic mulch and its inappropriate disposal, together with mechanized tillage practices, can generate a large number of macro- and microplastics that are incorporated into agricultural soils [33].

Another possible explanation may be attributed to wind dispersal, as there is evidence that plastic debris can be spread between different ecosystems and in other environments due to wind action [34].

Although biodegradable plastic mulch films have entered the sector as an alternative to traditional polyethylene mulch plastic, this supposedly more environmentally friendly type of film has also generated debates about its use [35]. The polymers used to produce biodegradable mulch films are not always fully renewable but can be a mixture of bio-based components with non-renewable biodegradable components, as well as incorporating chemicals such as plasticizers, antioxidants, and pigments, which can have an environmental impact as they can be released from the plastics as they degrade. Even plastics with pro-oxidant additives (PAC), also known as oxo-degradable plastics, consist mostly of LDPE with additives to enhance abiotic degradation, but with low or no degradation by microorganisms [1].

In addition, pesticides in contact with mulch plastics take longer to degrade. A team of researchers from the universities of Wageningen and Polytechnic of Cartagena participating in the European project *Diverfarming* have analyzed the interaction of these plastics (LDPE, PAC, and Bio) with 38 pesticides, where the research has shown how biodegradable plastics have a higher percentage of pesticide sorption. Specifically, the percentage of sorption was 23% in LDPE and PAC and 50% in Bio. In addition, the decomposition of the active substances in the presence of plastic was 30% lower than in the absence of plastic. This would imply that the microorganisms in charge of degrading these biodegradable plastics could be contaminated and the efficiency of the pesticides would also be compromised as the substances remaining in the plastics are not available in the soil to perform their function [36].

Concerning the management of agricultural plastics, the desired alternative to landfill disposal and incineration of mulch plastics is recycling, provided they are collected clean and separated. However, the lack of recycling capacity, the large volume of waste, and the low value of post-consumer plastics (PCr) have created great difficulties for the recycling industry and have resulted in limited options for

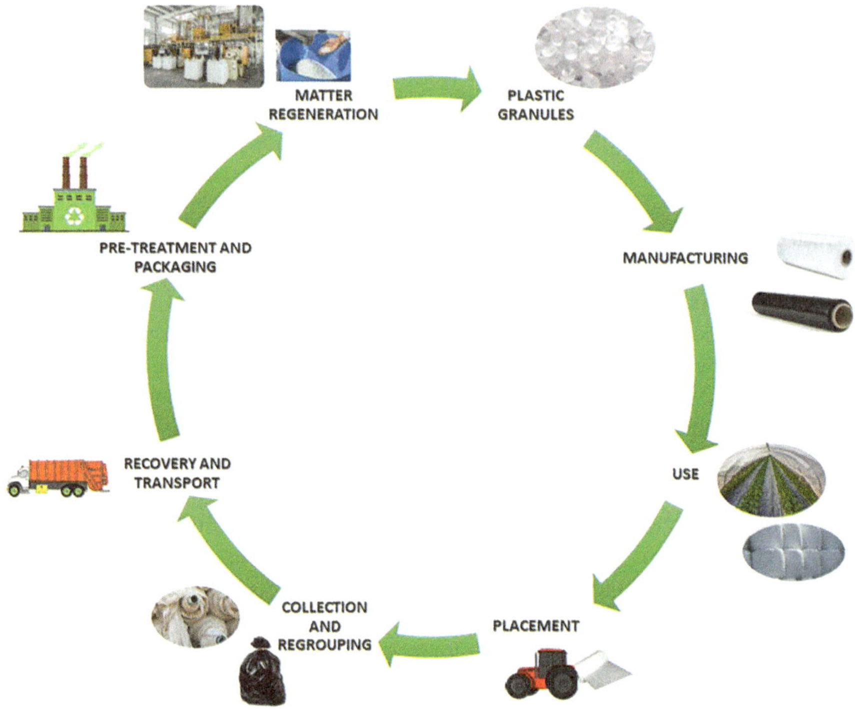

Fig. 4 Plastics in ecologically intensive agriculture and the circular economy in nine steps. Modified to [30]

collecting agricultural plastic waste. Chemical recycling could be an interesting option for the future, once agricultural plastics waste is compatible [30].

In this context, on December 2, 2020, the APE Europe General Assembly approved the European Plasticulture Strategy, a commitment to address the environmental challenges of the use of plastics in agriculture and to provide a homogeneous model for the sustainable management of plastics at the end of their useful life across Europe [37].

The European plasticulture strategy consists of recommending technical and economic systems and actions for the management of agricultural plastics and the reduction of their impact on the environment, through individual and collective initiatives involving farmers and livestock farmers, distributors, and plastic processors. This Circular Economy model is summarized in Fig. 4.

In addition, EPRO (European Association of Plastics Recycling and Recovery Organisations), made up of 19 organizations from 15 countries working together to promote the recycling of plastics, is also involved in this initiative. EPRO aims to exchange experiences to optimize the effectiveness of national plastic recycling systems through international cooperation [38].

4 Presence of Microplastics in Soils

The presence of microplastics in the soil has different origins such as plastic mulch films, municipal solid waste, compost, biosolids (sewage sludge or anaerobic digestate), plastics for agriculture, and atmospheric depositions, among others [39–41]. Some of the main microplastic sources are briefly described below.

Plastic Mulching
As indicated before, plastic mulching composed of polyvinyl chloride (PVC) and polyethylene (PE) mainly is a technique that is being used in agricultural production in many regions of the world due to its good properties. The plastic mulches provide thermal insulation, regulate the consumption of water and nutrient resources, improve the production of plant species, and provide economic benefits [40, 42, 43].

Despite these recognized benefits of plastic mulching, their use also represents a contamination problem. For example, the removal of mulching film from farmlands at the end of growing seasons can be difficult, which is why remnants of these films are often left on the farmland. In addition, these plastics degrade helped by exposure to the sun, rain, or the work of the crop itself, which favors the formation of micro- and nanoplastics. Some research has shown that plastic film mulches are one of the major microplastic contaminants on farmland ([32]; [41]; [44]).

Biosolids and Irrigation Water
The wastewater treatment plants receive wastewater with microplastics from different routes (cleaning and personal care products, textile washing, tire wear, etc.). These microplastics are usually recovered in the sludge fraction of treatment plants [41].

These biosolids are commonly used as fertilizers in agriculture due to their content of organic matter and nutrients. However, it has also been shown that biosolids can contain heavy metals and other contaminants, as well as microplastics. Therefore, the continued use of biosolids can contribute to the increasing presence of microplastics in the soil [40, 41, 43].

Furthermore, although most of the microplastics contained in wastewater are removed in the treatment process, significant concentrations of microplastics have been found in this wastewater ([43]; [45]). When this treated water is used as irrigation water, the microplastics will be transferred to the soil.

Compost from Municipal Solid Waste
The use of compost in agriculture is another way of introducing microplastics into soils. Compost from municipal solid waste treatment plants is used as an organic amendment and may contain microplastics. The plastic content in compost is caused mainly when the organic matter has not been selectively collected, being accompanied by bags and other plastic materials ([46]; [47]; [48]). Although the presence of microplastics in compost may vary depending on how this compost has been obtained, some research has confirmed that the use of this type of fertilizer is a way for microplastics to enter the soil [49].

Atmospheric Deposition
Atmospheric deposition is an important way of contributing to the presence of microplastics in the soil. The atmosphere is a medium in which various suspended

materials as microplastics can be transported from one place to another, with numerous studies confirming this fact. Zhang et al. [50] studied the presence of microplastics in the atmosphere, including detection, collection, and analysis methods. They also included the main aspects related to the characteristics of microplastics present in the atmosphere such as abundance, size, color, and polymer. Allen et al. [51] focused on the transport of microplastics in the atmosphere and their deposition in a remote area of the French Pyrenees, showing that microplastics can be transported long distances through the atmosphere. Liu et al. [52] studied the presence of microplastics suspended in the atmosphere in Shanghai, finding that approximately 120.7 kg of microplastics is transported per year through the air.

However, it must be taken into account that it is difficult to establish the ways by which these microplastics can be transported since aspects such as wind and its movement, currents, turbulence, geographical accidents, and the degradation that plastics can suffer from friction, ultraviolet radiation, and other environmental phenomena will influence the transport and subsequent deposition of microplastics in different environments. Therefore, detailed and complex studies are necessary to be able to establish the origin of the microplastics deposited in the soil.

The presence of microplastics in aqueous media has been extensively studied in recent years. However, research on microplastics in soils is even less developed, requiring more studies to evaluate the sources and pathways for the presence of microplastics in soils, the potential risks to the environment, and ways to avoid and control the presence of microplastics in soils. Table 2 shows a summary of recent studies investigating the presence of microplastics in soils.

5 Effects of Plastic Film Mulching in the Soil Environment of Agricultural Lands

As it was mentioned in the previous sections, plastic film mulching has been widely used in agricultural production to improve the condition of the soil, increase crop yield, stimulate microbial activity, minimize the use of nitrogen fertilizers, warm the soil, and increase water-use efficiency, between others [59–65]. However, the repeated application of plastic mulches is also provoking some important problems resulting from their slow rate of degradation and incomplete recovery. As a consequence, large amounts of plastic film waste are generated and remained in the soils of agricultural lands [66, 67]. These plastic residues in these agricultural lands affect the physicochemical properties of the soil [66, 67]. Furthermore, when exposed to environmental conditions, a certain proportion of microplastics are generated and transferred to animals or land and water bodies provoking an accumulation into the environment of a complex mixture of plastic fragments [68].

Some authors have investigated the effect of plastic film mulching on the distribution of plastic residues in agricultural fields and have included references to the benefits or damages to the soil environment (including soil organisms). For example, recently Kumar and Merline Sheela [69] identified plastic residues in the following four regions of Tamil Nadu, India: Sulur, Nagondapalli, Royakottah, and

Table 2 Recent studies on microplastics in soils

Studied region	MPs source	Study conditions	Main results	Reference
Chile's Metropolitan Region	Agriculture; mining; roadways; urban environment	Centrifuge and filtration of the supernatant; identification by optical microscope and FTIR microscope; statistical analysis: Kruskal-Wallis	Higher concentration of MPs on cropland and pastures Urban areas and roads did not influence the number of microplastics Fiber was the most common form of microplastic, followed by films. Acrylates, polyurethane, and varnish were the most common plastic polymers, followed by polyethylene	[47])
Tainan City (Taiwan)	Greenhouse; organic fertilizer; residential area; plastic mulch; protective foam	Dry and sieve; density separation with saturated NaCl solution; filtration; Fenton's reagent; identification by stereomicroscope; polymer composition of MPs by FTIR	High concentration of MPs on land adjacent to suburban roads due to fragmentation and sunlight exposure of protective fruit foams, and also due to particles generated by friction between the tire and the road On cropland, plastic mulch is an important source of MPs due to photolytic, mechanical, and biological degradation	[42])
Murcia (Spain)	Plastic mulch	Centrifuge and filtration of the supernatant; identification by stereomicroscope; statistical analysis: Shapiro-Wilk	Ingestion of microplastics in livestock. Results show that 100% of the soil samples analyzed contained MPs, as well as 92% of the sheep feces samples studied	[53])
Shihézi (China)	Plastic mulching with different ages (5, 15, and 24 years)	Stirring and density separation with NaI; samples digestion with H_2O_2; identification by light microscopy and FTIR	The higher the age, the higher the concentration of MPs The main composition found was PE and the predominant shape was a fragment	[32])

(continued)

Table 2 (continued)

Studied region	MPs source	Study conditions	Main results	Reference
Yucatan Peninsula (Mexico)	Horticulture; lit	Ultrasonication and density separation in water; identification by light microscopy	MPs concentrations increased from soil (0.87 items/kg) to earthworm and chicken feces, making a point on the exposure risk for animals and humans 60% of the MPs in soils were below 20 μm in size	[54])
Malmö (Sweden)	Agriculture; sewage sludge	Treatment with H_2O_2 + enzymes and density separation with $ZnCl_2$; identification by light microscopy, FTIR, and shape-to-mass	MPs concentration in soils varied from those receiving sludge as fertilizer or not. Soils that did not receive sludge achieved higher MPs concentrations than those receiving 1 ton/year, whereas soils receiving 3 tons/year showed MPs concentrations 10 times higher	[55])
Fars province (Iran)	Plastic mulching	Stirring, ultrasonication, and density separation in water; identification by light microscopy and shape-to-mass	The concentration of light microplastics was below 7 items/kg The composition of MPs was not possible to identify. The sizes were considerably lower than those found in other studies Wind erosion plays a key role in the spreading of MPs	[56])
Southeast Germany	Plastic mulching, greenhouses, ensilage	Treatment with H_2O_2 and sieving; identification by stereomicroscope and ATR-FTIR	The authors found 206 macroplastic pieces per hectare and 0.34 ± 0.36 microplastic particles per kilogram dry weight of soil Polyethylene was the most common polymer type, followed by polystyrene and polypropylene	[44])

(continued)

Table 2 (continued)

Studied region	MPs source	Study conditions	Main results	Reference
			Films and fragments were the dominating categories found for microplastics, whereas predominantly films were found for macroplastics	
Ilia County, Western Greece	Plastic mulching	Sieving and density separation with NaCl; samples digestion with H_2O_2; identification by stereomicroscope, ATR-FTIR, and SEM	The microplastics found were black and derived from the plastic mulch used in the cultivation of watermelons and canning tomatoes The MPs were oxidized by their exposure to sunlight and were broken down into smaller pieces with edges The concentration of MPs was higher in watermelon (301 items/kg) than in tomatoes (69 items/kg)	[57])
Hangzhou Bay, East China	Plastic mulching in cropped lands	Continuous-flow, air floating device and density separation with NaCl and NaI in the second step; identification by stereomicroscope and μ-FTIR	Mulching soil contained more MPs than non-mulching soils (571 vs. 263 items/kg, respectively). Fibers were the most abundant shape, with multiple compositions: PE, PP, polyester, rayon, etc.	[58])

Krishnagiri. Their results indicated that the dehydrogenase enzyme activity of the rhizosphere soil was not affected by the mulching practice. However, a long-term study is required to present stronger conclusions about the effect of plastic film mulching on the microbial community structure and other micro- and macro-fauna in soil with mulching. These authors also confirmed that the duration of mulching plays an important role in the distribution of plastic residues. Furthermore, Li et al. [70] reported that PE plastic mulch is the main source of microplastics in the farmlands of Xinjiang, China, and found that the number of microplastics increased as the continuous mulch period increased (from 5 to 30 years) in the range of 10.10–61.05 mg/kg, respectively. However, these authors did not report the implications for crops or soil microbiota. Other authors specifically explored how the presence of microplastics from mulching film affects soil microbial communities [67]. Their results demonstrated that mulching film microplastics may be a "special microbial accumulator," enriching the bacterial groups involved in their biodegradation. Also, Zhu et al. [71] reported that the intestinal flora of *Folsomia candida* was changed due to microplastic pollution in soil, and Maab et al. [72] showed that microplastic particles can be moved and distributed by soil microarthropods. Some other additional studies have also reported the positive and negative effects of microplastics on soil organisms [73–75]. However, in general, these works are subjected to limitations and future research will need to better elucidate the impact of plastic mulching microplastics on soil biota. Regarding effects on water distribution and flow processes in the soil, Jiang et al. [76] reported that the presence of residual plastic film fragments significantly altered soil water distribution causing heterogeneous water infiltration.

Special attention has been focused on plastic mulch additives that are easily released into the soil and that can cause adverse effects on soil organisms [77, 78]. The toxicity of these chemicals associated with microplastic pollution should be taken into account and carefully evaluated [79]. The case of phthalate esters (PAE) has been specially analyzed [80–82]. Polyethylene (PE) is the main type of shredded plastic film used in agriculture, and some studies have found that PAE concentrations in PE film are in the order of tens to hundreds of mg/kg [83, 84]. Consequently, high levels of PAEs have been reported to accumulate in agricultural fields under continuous plastic mulching, as well as their subsequent absorption and accumulation in cultivated plants [83, 85, 86]. The main observed effect associated with these PAE concentrations was inhibition of plant development in several cultivated species [87–89]. In this sense, the recent study of [90]) investigated the effects of PE film on PAE concentrations in the soil-crop system and evaluated the risks for people who consume agricultural products. The concentration of PAE in the plastic-covered agricultural soil layer was 0.45–0.81 mg/kg, while the average concentration in uncovered soils was 0.37–0.73 mg/kg (19% lower). On the other hand, the PAE concentration in mulched soil initially increased and then decreased (from May to October) and reached its highest level in July, being 0.80–0.84 mg/kg. However, in the soil of non-mulched farmland, PAE concentration decreased gradually over time. On the other hand, the authors found that the roots of corn (*Zea mays L.*) could absorb and accumulate PAE, and the bioconcentration factor was 1.6–2.3, and the average concentrations of PAE in stems, leaves, and grains were 0.77–1.47 mg/kg.

In addition to the release of plastic mulch additives, some toxic compounds such as heavy metals or persistent organic pollutants (POPs) can be adsorbed by generated microplastics and transported into the food chain. These sorbed chemicals are of particular concern since they can cause severe toxic effects [91]. For example, Ramos et al. [92] studied the pesticide release from polyethylene film (previously adsorbed to the plastic surface) and found that the pesticides accumulated in the plastic film can easily migrate to soil and atmosphere.

Regarding effects on fauna, there is only one study that reported the presence of microplastics from plastic mulching on animals. This interesting work was carried out in an intensive area of plastic mulching in Murcia (Spain) by Beriot et al. [53]. In the Murcia region, plastic mulching is used mainly for the cultivation of vegetables, and after the collection, sheep graze on the vegetable residues. The authors analyzed microplastics in both soil and sheep. Beriot et al. [53] found $2 \cdot 10^3$ particles/kg in the soil and 10^3 particles/kg in the sheep feces. These results demonstrated that plastic particles are in the soil but also that sheep ingested them. This first work of plastic mulch in fauna in Spain suggests that further studies should be oriented to estimate the effects of this plastic waste on the health of flora, fauna, and humans. Although it is not a specific study of plastic mulching, Lwanga et al. [93] also studied the increase in the concentration of microplastics in home garden soil by analyzing earthworm casts and chicken feces. The concentration of microplastics was 129.8 ± 82.3 particles/g in chicken feces and 10.2 ± 13.8 plastic particles per gizzard of the chicken. Due to data shortage and the important risk to animals and human health, the potential effects of plastic film mulching on soil (also including microorganisms and animals) should be further examined with future research studies.

Finally, some other interesting studies explored the impacts of micro- and nanoplastics on plants [94, 95] or soil nutrient cycling [96]. Although the effect on soil nutrient cycling is still unknown, [96]) identified several potential ecological risks associated with N cycling and surprisingly found that biodegradable plastic was a more potent inhibitor of plant growth and development. Also, [94]) reported that cucumber plants' growth was significantly decreased after exposure to polystyrene particles. In addition, Lwanga et al. [95] found that earthworm mortality was increased due to exposure to microplastic pollution.

To conclude, the lack of studies to analyze the life cycle of plastic mulching and provide efficient systems to collect, recover, and recycle agricultural plastic waste has led to the disposal of large quantities of used plastics in the environment. Unfortunately, in some cases, the cost and effort of collection, reuse, and recycling operations seem to outweigh their benefits. A detailed and comparative assessment of the full life cycle could make the use of plastic mulching environmentally unsustainable and highlight the potential risks of plastic mulching [65].

6 Sustainable Alternatives to Plastic Mulch

Numerous sustainable alternatives to plastic mulch can be used, such as mulch made of cellulose, straw and other organic materials, wood, wool, and even "living" mulches, usually composed of clover covers [97–99]. In this section, the main characteristics of some alternative mulches will be described. However, the most widely studied alternatives are currently biodegradable mulch plastics, designed as substitutes for the traditional PE that makes up these agricultural plastics, which are buried in the soil after use, where they are degraded by the microorganisms present [91].

6.1 Biodegradable Plastics

Examples of biodegradable plastics widely used in other sectors are polylactic acid (PLA) from starch-corn blends, polyhydroxyalkanoates (PHA), or polybutylene succinate (PBS) [100]. There are also biodegradable plastics based on fossil resources, such as poly (butylene adipate-co-terephthalate) (PBAT). All polymers used for the manufacture of biodegradable plastics contain ester bonds or are polysaccharides, which makes them susceptible to microbial hydrolysis and allows them to be converted into biomass, CO_2, and water [101]. Several studies have established these facts through theoretical tests, but the breakdown of biodegradable plastic within a reasonable time does not always occur [91].

Advantages of Biodegradable Mulches
Some of the advantages of biodegradable mulch plastics over traditional ones are mainly related to soil microclimatic variations. In contrast to the large greenhouse effect generated by traditional PE mulch on the soil, biodegradable mulches generate less heat and are favorable in tomato crops in areas characterized by high temperatures [102]. In another study conducted on vineyards in Mediterranean climate [103] with mulch films based on PBAT, they obtained a lower vapor permeability, favoring fruit yield and root development and exudation, due to the loss of moisture from the topsoil after a few months of trials. This root development also leads to a greater amount of nutrients available to rhizophore microorganisms, increasing the richness of the upper soil horizon [104].

Biodegradable plastics also have a positive aspect at the end of their useful life, as they are prepared to be buried in the soil, and this limits uncontrolled dumping and burning of plastic waste, and also the possibility of it ending up in the sea [91, 105]. In addition, they can generate positive effects on soil and microbial communities when they are buried for biodegradation. This burial means an important change for the soil from a physical and biogeochemical point of view, but it is still unclear what effects it produces and how it differs from traditional mulching. From a toxicological point of view, biodegradable plastics incorporated into the soil are considered environmentally safe. For example, tests on biodegradable plastics blended with copolyester and starch have shown no adverse effects on bacterial reproduction or soil nitrification, although these are short-term studies [106, 107].

Buried plastics also contribute significant carbon to the soil, affecting the development and behavior of some microbial communities, although these effects are not always negative. For example, an increase in the enzymatic activity of some microbes and an improvement in the development of species such as *Aspergillus*, *Penicillium,* and the fungus *Acanthamoeba* has been observed ([102]; [108]) in the presence of PBSA (poly (butylene succinate-co-butylene adipate) bioplastic), while other Ascomycota and Proteobacteria species are favored by the use of PBAT plastics [109]. However, most of these studies only examine one soil type or geographical location, so it should be taken into account that variations in microbial communities may also be due to natural variations in soil composition, external climatic changes, and natural changes in the microbial communities themselves [91].

How Does the Biodegradation Process Occur?

For a polymer to be biodegradable, the extracellular enzymes of the microorganisms have to break down the polymers in the chains into monomers, which are then used for bacterial growth. This can result in the mineralization of the polymer molecules to their basic compounds, CO_2 or CH_4 depending on aerobic or anaerobic conditions, H_2O, and minerals, increasing biomass and leaving no plastic residues in the soil [110].

To achieve biodegradation, the presence of hydrolyzable bonds between the polymer monomers is required, which will be used by microorganisms to grow once they are detached from the polymer chain and incorporated into the soil. The biodegradation process comprises three steps [110]: (1) microbial colonization of the polymer surface, mainly by bacteria and fungi; (2) depolymerization by the extracellular enzymes of the microorganisms; and (3) utilization of the hydrolysis products by the microorganisms for growth.

Some studies have demonstrated the successful biodegradability of certain types of plastics, such as PBAT, over a 6-week incubation period [111]. In this study, the colonization of PBAT by fungi generated highly enriched carbon compounds that were stored in the soil and subsequently utilized by the fungi when the availability of other nutrients was limited. The main factors affecting the biodegradation process are as follows [112].

- Abiotic environmental factors:
  - Moisture: the rate of polymer degradation is increased in the presence of enough moisture to allow adequate growth and multiplication of bacteria. In addition, highly humid areas favor the hydrolysis process.
  - pH: this factor has a twofold effect. On the one hand, it can alter the rate of hydrolysis reactions (at pH 5 PLA hydrolysis is optimal, for example), while the degradation products of bioplastics can alter the pH of the soil and thus bacterial growth.
  - Temperature: the degradability of bioplastics decreases with increasing soil temperature. Moreover, the higher the melting point of the polymers, the lower the rate of biodegradation.
- Biotic environmental factors:
  - Enzymes characteristics: depolymerize to a greater or lesser extent depending on the active sites the enzymes have on their surface to act on different types of monomers.
- Polymer factors:
  - Molecular weight: the higher the molecular weight, the lower the biodegradability.
  - Shape and size: polymers with a high specific surface area can be degraded more rapidly than others.
  - Additives: some additives added to bioplastics, such as cellulosic filler, reduce the thermal stability of the bioplastics, and increase the ash content. Also, metals added as antioxidants can be sensitive to thermo-oxidative degradation.

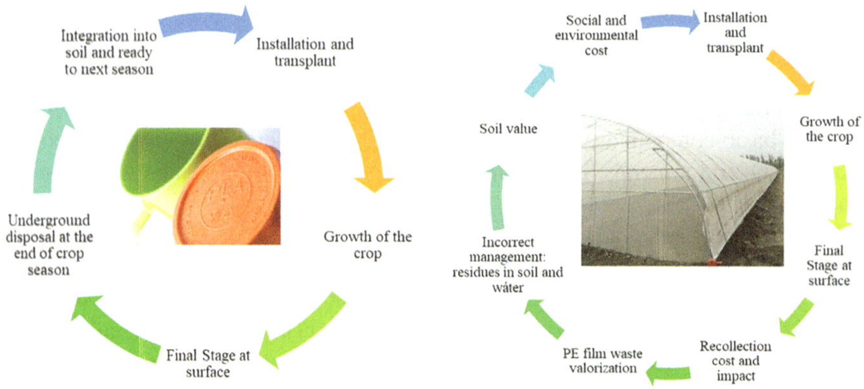

Fig. 5 The life cycle of biodegradable plastics vs conventional plastics. Adapted from [114]

Potential Risks Derived from the Use of Bioplastics in Agriculture

One of the main problems currently facing biodegradable plastics is the general lack of knowledge about their characteristics, especially for farmers. A study conducted in three US states during the years 2008–2012 found interesting conclusions [113]. Firstly, the information available to citizens about the characteristics of these plastics was still very deficient. The lack of knowledge about the advantages of these types of plastics means that the user does not buy them and continues to opt for the traditional version. On the other hand, the cost of biodegradable plastics is significantly higher than that of PE, while the waste management costs are much higher for PE and consist of several management steps that biodegradable plastic does not require (Fig. 5) [114].

In addition, there is evidence of certain toxicological risks that may occur when bioplastics are not properly managed after their useful life and are abandoned in the environment. Recently, a review of bioplastics has shown that microplastics from biodegradable plastics can have harmful effects on certain organisms [115]. Mulch plastics are ready to be buried in the soil after use, but if they are not properly managed, they can also end up affecting the environment. For example, high concentrations of PLA can cause respiratory disorders in marine species such as *Ostrea Edulis* or *Arenicola marina L.* [116, 117]. Some studies show the capacity of bioplastics to adsorb organic pollutants. Specifically, Zuo et al. [118] studied the adsorption of phenanthrene on the biodegradable plastic PBAT versus two conventional plastics, PE and PS. The authors found that the adsorption and desorption capacity of this pollutant on PBAT was higher than on conventional plastics. On the other hand, very few studies have focused on the ecotoxicity of bioplastics when degraded in the presence of organisms other than bacteria. Recently, González-Pleiter et al. [119] studied the toxic effects of PHB-derived nanoplastics on two types of algae and the crustacean *Daphnia Magna*. The authors found a decrease in cell growth and alteration of physiological parameters of the algae, while a decreased rate of movement and membrane damage was observed in the crustacean. This is just one example that raises awareness that biodegradable plastics are a good alternative to traditional plastics, but there is an environmental risk that is increased when waste is improperly managed.

Finally, additives are often added to biodegradable plastics to improve the mechanical properties of the material and to achieve good functional efficiency during the growing cycle. Additives are also added to promote the biodegradation of the plastic in the soil. All these substances are likely to be released into the soil when the plastic is buried for degradation [120]. The ecotoxicity of the substances released by bioplastics into the soil has been little studied, and only a few publications have focused on their effect on certain soil microorganisms. For example, Sforzini et al. [107] conducted a laboratory study in which they subjected a biodegradable plastic to degradation for 6 months in the presence of various bacteria, protozoa, green algae, monocotyledonous plants, and invertebrate animals. The results showed that this plastic was safe for agricultural use and no toxicological effects were observed in the short to medium term. However, other studies on the effect of small plastic fragments buried in the soil on crops planted in that area (wheat, tomato, lettuce, or maize, among others) have shown, in general, lower efficiency of crop germination, plant biomass, and overall growth [35, 120–122].

Therefore, biodegradable plastics are a good alternative to traditional PE mulch, as long as they are properly managed after their useful life and the soil in which they are buried is used deliberately. The main types of biodegradable plastics used in agriculture, their origin, and their biodegradability in soils are summarized in Table 3.

6.2 *Paper Mulch*

Another more sustainable alternative to traditional PE mulch plastic is paper mulch. It is used in mulch lower proportion than biodegradable plastics, although it can be a cheaper option depending on the thickness of the mulch (Fig. 6) [124].

Paper mulch usually consists of an agglomerate of wet cellulosic fibers derived from wood, which are subsequently dried and held together by hydrogen bonds. The composition of these fibers is usually 40–45% cellulose, 25–35% hemicellulose, and another 25–35% lignin. The length and diameter of the fibers are usually around 1–3 mm and 0.009–0.03 mm, respectively [124]. Normally, this material is porous and hygroscopic and can expand and shrink with changes in moisture content, which is also influential when it starts to biodegrade in soil. As biodegradable plastics, paper mulch is buried in the soil at the end of its life and biodegraded there. However, depending on the composition the effects on the soil and the microorganisms are different.

The advantages of this type of mulch depend mainly on the type of crop on which it is applied. Some studies claim that with paper mulch, plants need more irrigation because paper retains less moisture than plastic [124]. In addition, paper mulch may encourage weed growth as it is more transparent to sunlight than traditional plastic.

Table 3 Main biodegradable plastics used in agriculture, origin, and biodegradability in soil. Adapted from Serrano-Ruiz et al. [110]

Compound	Origin	Biodegradability in soil
Polylactic acid (PLA) 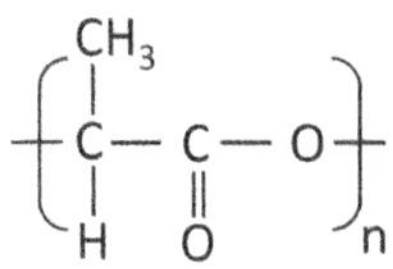	Natural or synthetic (blend of starch and corn)	Low (better in combination with others)
Poly (hydroxyalkanoates) (PHA) 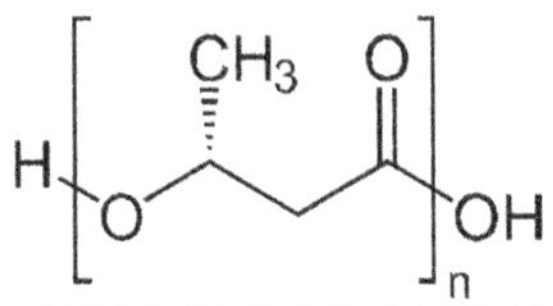	Natural or synthetic (from some bacteria and fatty acids)	Low (better in combination with PLA and others)
Poly (butylene succinate) (PBS) 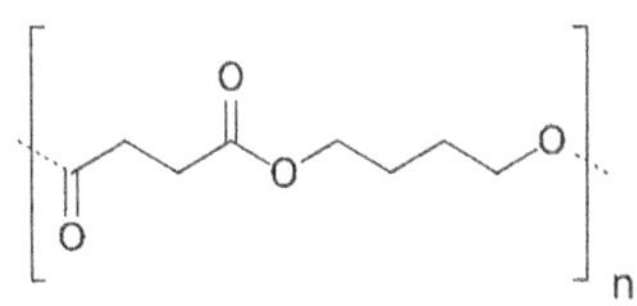	Synthetic (from succinic acid)	Moderate (better in combination with PLA and PBSA)
Poly (butylene succinate adipate) (PBSA) 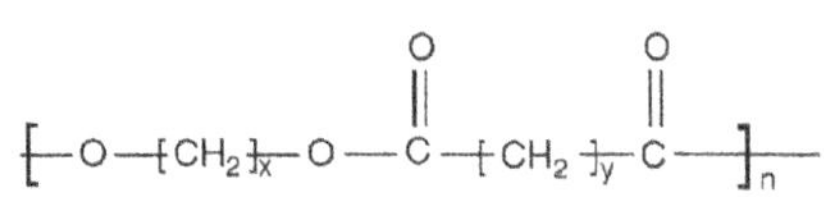	Synthetic (from succinic acid)	Moderate (better in combination with PLA and PBS)
Poly (butylene-adipate terephthalate) (PBAT) 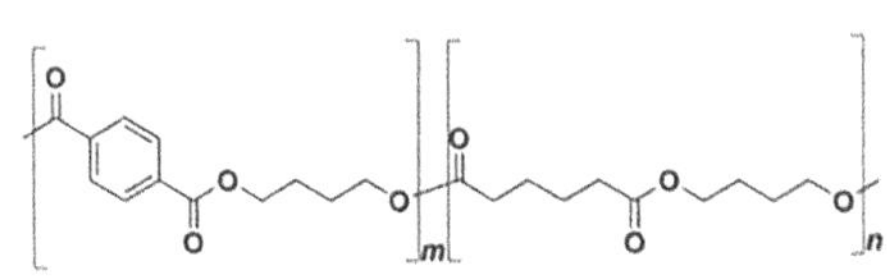	Synthetic (from butanediol and adipic acid)	Moderately low (better in combination with PLA or starch)
Poly (3-hydroxy butyrate) (PHB) 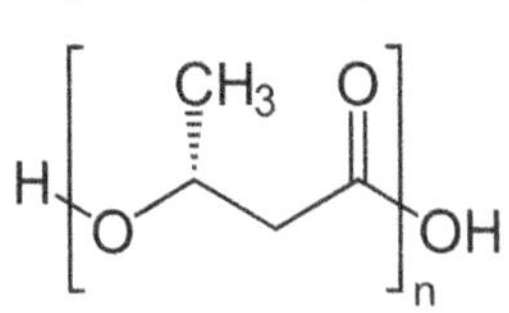	Natural (from bacteria)	Moderate (better in combination with TiO_2 composites)
Poly (caprolactone) (PCL) 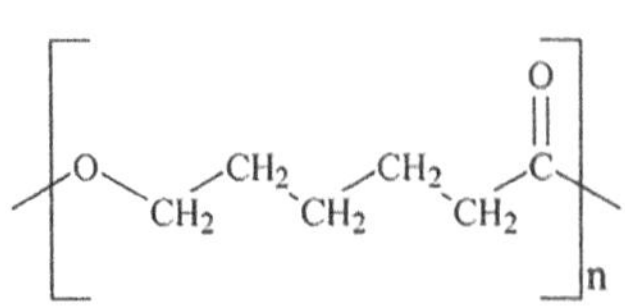	Synthetic (from ε-caprolactone)	Moderate (better in combination with PLA or PHB)

Fig. 6 Example of a paper mulch used in agriculture. Source: HSMG [123]

However, this can be overcome by choosing mulch treated with waxes and black pigments on one side, which makes it more impermeable. Other advantages are that paper keeps the temperature of the soil cooler, as the lighter color reflects sunlight. Therefore, it can be a good alternative to use in cold places or seasons where crops do not require soil warmth [124]. Contrary to Haapala et al. [124], other studies have shown that recycled paper mulch can decrease the water consumption needed by some crops, as it reduces evaporation by restricting the flow of water and vapor through the soil–atmosphere interface, which causes roots to retain water longer, proving to be a suitable alternative to polyethylene [124, 125].

One of the main disadvantages of this material is that, during the growing cycle, it undergoes uneven degradation, as the parts that are covered by the soil degrade first and then the visible part starts to degrade. The quality and type of fibers of which the paper is composed have a significant effect on degradability and the time it takes. For example, one study found that mulch made from shredded newspaper resulted in uneven degradation leading to the generation of weeds in experimental plots [126]. Other studies have shown that using recycled paper to make this type of mulch caused degradation within 40–50 days of planting. In other words, in general, paper mulch degrades more rapidly than biodegradable plastic mulch and can start to degrade in areas in contact with the soil even before the end of its useful life.

The degradation rate of paper mulch is controlled by the lignin content of the fibers. The lower the lignin content, the shorter the time it takes to biodegrade [127]. Additives are sometimes added to increase the biodegradation time of paper, but these additives must meet many requirements: they must be abundant and accessible to paper manufacturers, environmentally safe, and low cost. The main additives added to paper mulch are lignin sulfate (Indulin AT), acid minerals, and

Table 4 Main types of paper mulch, durability, and effectiveness in weed control. Source: Adapted from Ahokas et al. [127]

Compound	Durability	Weed control	Effect on yield[a]
Kraft paper 90 g/m^2	Degradation of buried edges during the cropping period	Effective	Positive
Kraft paper 200 g/m^2	Some wind damage	Effective	Positive
Black kraft paper 83 g/m^2 + latex	n.d.	Effective	Positive
Beige kraft paper 83 g/m^2 + biodegradable plastic	n.d.	Poor/moderate	Positive
Kraft paper with black pigment	Blown off most plots	n.d.	Positive
Brown kraft paper	n.d.	Effective	Positive
Brown kraft paper + linseed oil + Co + carbon black	Good	Effective	Positive
10 μm white paper + biodegradable plastic	n.d.	Moderate	Positive
10 μm black paper + biodegradable plastic	n.d.	Effective	Positive
Asphalt-saturated paper	n.d.	Effective	Positive
Shredded newspaper, different depths	n.d.	More effective with increasing depth values	Positive
Shredded paper glued between two sheets of brown paper	Dislodged by wind and exposed sites	Effective	Positive
Paper derived from recycled fibers	Some wind damage	Moderate	Negative

[a]Compared to bare soil

formats [127]. These are natural additives, less harmful than those contained in traditional plastics.

In general, there seems to be a growing interest in using paper mulch on certain types of crops because of its good biodegradability and because it is made of materials from renewable sources. The main types of paper mulch that can be used, their durability, and their effectiveness against weed control are shown in Table 4.

6.3 *Straw Mulch*

Straw mulch is another of the most universally used options in agriculture. It has comparable benefits to traditional plastic but is much more environmentally sustainable than traditional plastic. From a crop perspective, it prevents weed growth, reduces fertilizer leaching, and promotes soil moisture retention (Fig. 7) [129].

Fig. 7 Rice straw mulching in potato. Source: [128])

It can also protect crops from the direct impact of heavy rainfall better than plastic, which is of interest in tropical areas and areas susceptible to heavy rainfall [130]. Soil moisture is retained because straw reduces turbulent air exchange between the soil and the atmosphere. It is also a useful material in rainfed crops, as it can reduce latent heat flow from the soil and reduce evaporation between wheat plants in winter, allowing more water to accumulate in the soil.

Straw mulching has become an ideal supplement for soil nutrients and increasing crop yields in modern agriculture, as it is rich in nutrients and physiologically active substances. It is of great ecological importance for maintaining crop fertility, reducing the use of chemical fertilizers, improving the soil's capacity as a carbon sink, and also promoting nitrogen cycling in the soil [131]. It is a type of mulching that can be used not only in intensive crops but has proven to have good applications in orchards where the soil suffers from severe soil erosion and loss of fertility [131].

The main drawback of straw mulch is that it needs to be applied in fairly accurate amounts to avoid affecting crop yields. To illustrate this effect, a study carried out on wheat crops covered with straw mulch showed that mulching at a rate of 2000–4000 kg/ha increased maize growth while mulching at a rate of 6000–8000 kg/ha significantly decreased maize yield [132]. This effect can also be contrasted in other studies [133] which show that a rate of 4000–4500 kg/ha is sufficient to achieve good results in the first year of the farm and can be reduced in the following years. Also, the study by Peng et al. [130] showed that the application of straw mulch during the crop growth stage improved crop yields.

6.4 Wool Mulch

The use of this type of mulch began in 1998 when the University of Minnesota began collaborating with the Minnesota Wool and Lamb Producers Association (MLWPA) and the Agricultural Utilization Research Institute (AURI) to find an alternative use for the large amount of wool produced in the state after the shearing season. The value of wool was very low at the time and the university began to research into the

Fig. 8 Example of wool mulch used in agriculture. Source: [98])

use of biological mulches for crops that would control weed generation and reduce reliance on pesticides [98]. Finally, in 2004, the research bore fruit and experiments led to the creation of a mixture of wool fibers and wood shavings that provided a lightweight and economically viable material that was effective in suppressing weeds, retained soil moisture, and stabilized soil temperature (Fig. 8) [94]. Other studies have demonstrated the effectiveness of this material for weed control in laboratory trials with natural wool mulches treated with copper hydroxide, isoxaben, or prodiamine [134].

6.5 "Living" Mulches

It is a crop cover method based on certain perennial plant species that emerged in response to the need to control weed growth in agriculture. The ability of these mulches to control weed production can vary from 34% to 96% [135]. Traditionally, white clover has been the most widely used type of living mulch by farmers. It is highly adaptive and can provide suitable habitat for beneficial insects in crops, as well as being able to fix nitrogen in the soil [129]. The main drawback is that it tends to spread rapidly, so it is necessary to control its growth by mowing before flowering.

In addition to white clover mulches, there are also alternatives such as winter rye or fescue. These are grasses and legumes that are more winter hardy, improve soil cover, and are tolerant to climatic changes.

The positive effect of living mulches is therefore not limited to the control of weed growth. They are also favorable for limiting soil erosion, improving soil structure and biological activity, preventing nutrient leaching to lower horizons, limiting evaporation, and controlling soil and crop temperature [135].

The main disadvantage of this kind of mulch is that productivity may be reduced when using them, as they compete to some extent with the crop for light, water, and nutrients. However, some studies state that the success of using this type of mulch depends on the choice of the plant according to the type of crop, the time of planting, and the correct maintenance to limit the overgrowth of the mulch [136].

6.6 Economic Costs Comparison

Throughout this section, the main sustainable alternatives to the traditional plastic mulch used in agriculture are described. An important aspect to take into account is the economic cost of each of these materials, which will also determine whether or not they are chosen by the farmer. On many occasions, it would be necessary for farmers and people involved in intensive crops to receive training on these alternatives aimed not only at environmental improvement but also at improving the quality of the products as they are treated with more natural materials. This enhancement in productivity and quality can be reflected in an economic benefit for the producer, a point of view that should be emphasized so that people in the sector make an effort to invest in more sustainable materials than traditional plastic.

There are few economic studies available comparing the use of traditional PE versus biodegradable mulch or paper mulch, for example. In the European Union, the Agricultural European Innovation Partnership (EIP-AGRI) in its project "Reducing the plastic footprint of agriculture" does have a more or less comprehensive economic study comparing the full costs of using PE versus biodegradable plastics [137]. Indeed, the market prices of biodegradable films are higher than those of PE, which reduces their economic attractiveness for farmers in the short term. However, the total life cycle of these mulches must be considered to get a full insight into the costs.

For example, the cost of disposal of traditional PE is around 200–400 €/ha on average, whereas biodegradable plastics do not. Moreover, biodegradable plastics have a subsidy fund of up to 30% to encourage farmers to opt for these more sustainable options. With this support, the total cost/ha of using biodegradable plastics is similar to the cost of using PE. If the use of PE is considered without taking waste management into account, the total costs are 8700 €/ha, where mulching represents 6% of the total production costs. The highest expenditure for these operations corresponds to the growing season (45%) and the harvesting season (27%). In the case of using biodegradable mulch, the total cost would be 9000 €, with the materials representing 7–15% of the total costs. On the other hand, the use of these sustainable materials leads to a saving of 2% in the cultivation stage. In addition, the subsidy aid (35%) for the use of these materials has to be deducted.

These data are in line with those provided by the study by Marí et al. [138], which also includes an economic comparison with paper mulch. According to these authors, biodegradable mulch material is 25–188% more expensive than PE, while paper mulch materials are 153–236% more expensive (material cost alone). They

Table 5 Economic analysis and comparison of total agricultural costs using conventional plastic, biodegradable plastic, and paper mulch. Sources: Adapted from [137, 138])

Operations	Waste scenario	Costs (€/ha)
Field preparation		1448
Crop season operations		3931
Average conventional PE		548
Average biodegradable		987
• Mater-Bi®		1308
• Sphere®		916
• Ecovio®		649
• Bioflex®		1075
Average paper mulch		1300
• Mimgreen®		1264
• Arrosi® 69		1202
• Arrosi® G 1[a]		1536
• Arrosi® 240		1202
Harvest		2340
Field conditioning PE	No waste management	432
	Landfill	467
	Recycling	452
Field conditioning biodegradable plastic and paper		232
Total conventional PE		8676
Total biodegradable plastic		8938
Total paper mulch		9250

also report that, in the overall cost of operations, the use of paper mulch would account for 13–16% of the total, a higher percentage than that of biodegradable plastic.

When the waste management phase is taken into account, land preparation can cost up to 4.8% if the farmer transports the waste to the recycling point and up to 4.9% if he bears the full recycling cost. In contrast, the use of alternative mulch reduces land preparation costs by up to 57% compared to PE, from a waste management point of view. Table 5 shows the economic comparison between the different materials proposed in the two studies analyzed.

References

1. EIP-AGRI Focus Group (2021) Reducing the plastic footprint of agriculture. Final Report 2021, Brussels (Belgium). https://ec.europa.eu/eip/agriculture/en/publications/eip-agri-focus-group-plastic-footprint-final. Accessed 2 July 2021
2. Zhang H, Miles C, Gerdeman B, LaHue DG, DeVetter L (2021) Plastic mulch use in perennial fruit cropping systems – a review. Sci Hortic (Amsterdam) 281:109975. https://doi.org/10.1016/j.scienta.2021.109975

3. Agriculture Plastics Environment Europe (n.d.). https://apeeurope.eu/
4. Chen N, Li X, Šimůnek J, Shi H, Hu Q, Zhang Y (2021) Evaluating the effects of biodegradable and plastic film mulching on soil temperature in a drip-irrigated field. Soil Tillage Res 213:105116. https://doi.org/10.1016/j.still.2021.105116
5. Dai Z, Hu J, Fan J, Fu W, Wang H, Hao M (2021) No-tillage with mulching improves maize yield in dryland farming through regulating soil temperature, water and nitrate-N. Agric Ecosyst Environ 309:107288. https://doi.org/10.1016/j.agee.2020.107288
6. Zheng L, Pei J, Jin X, Schaeffer S, An T, Wang J (2018) Impact of plastic film mulching and fertilizers on the distribution of straw-derived nitrogen in a soil-plant system based on 15N–labeling. Geoderma 317:15–22. https://doi.org/10.1016/j.geoderma.2017.12.020
7. Bonanomi G, Chirico GB, Palladino M, Gaglione SA, Crispo DG, Lazzaro U, Sica B, Cesarano G, Ippolito F, Sarker TC, Rippa M, Scala F (2017) Combined application of photo-selective mulching films and beneficial microbes affects crop yield and irrigation water productivity in intensive farming systems. Agric Water Manag 184:104–113. https://doi.org/10.1016/j.agwat.2017.01.011
8. Lee ON, Park HY (2020) Effects of different colored film mulches on the growth and bolting time of radish (Raphanus sativus L.). Sci Hortic (Amsterdam) 266:109271. https://doi.org/10.1016/j.scienta.2020.109271
9. Jones H, Black TA, Jassal RS, Nesic Z, Johnson MS, Smukler S (2021) Characterization of shortwave and longwave properties of several plastic film mulches and their impact on the surface energy balance and soil temperature. Sol Energy 214:457–470. https://doi.org/10.1016/j.solener.2020.11.058
10. Solomakhin AA, Blanke MM (2007) Overcoming adverse effects of hailnets on fruit quality and microclimate in an apple orchard. J Sci Food Agric 87:2625–2637. https://doi.org/10.1002/jsfa.3022
11. Ibarra-Jiménez L, Lira-Saldivar RH, Valdez-Aguilar LA, Del Río JL (2011) Colored plastic mulches affect soil temperature and tuber production of potato. Acta Agric Scand Sect B Soil Plant Sci 61:365–371. https://doi.org/10.1080/09064710.2010.495724
12. Katan J, Gamliel A (2014) Plant health management: soil solarization. In: Encyclopedia of agriculture and food systems. Elsevier, pp 460–471. https://doi.org/10.1016/B978-0-444-52512-3.00256-4
13. Ding D, Wang N, Zhang X, Zou Y, Zhao Y, Xu Z, Chu X, Liu J, Bai Y, Feng S, Feng H, Siddique KHM, Wendroth O (2021) Quantifying the interaction of water and radiation use efficiency under plastic film mulch in winter wheat. Sci Total Environ 794:148704. https://doi.org/10.1016/j.scitotenv.2021.148704
14. Gao H, Yan C, Liu Q, Ding W, Chen B, Li Z (2019) Effects of plastic mulching and plastic residue on agricultural production: a meta-analysis. Sci Total Environ 651:484–492. https://doi.org/10.1016/j.scitotenv.2018.09.105
15. Wang Y, Li S, Cui Y, Qin S, Guo H, Yang D, Wang C (2021) Effect of drip irrigation on soil water balance and water use efficiency of maize in northwest China. Water 13:217. https://doi.org/10.3390/w13020217
16. Jabran K, Chauhan BS (2018) Weed control using ground cover systems. In: Non-chemical weed control. Elsevier, pp 61–71. https://doi.org/10.1016/B978-0-12-809881-3.00004-8
17. Kasirajan S, Ngouajio M (2012) Polyethylene and biodegradable mulches for agricultural applications: a review. Agron Sustain Dev 32:501–529. https://doi.org/10.1007/s13593-011-0068-3
18. PlascticsEurope Association of Plastics Manufacturers (2020) Plastics-the Facts 2020. An analysis of European plastics production, demand and waste data. https://www.plasticseurope.org/es/resources/market-data. Accessed 2 July 2021
19. Galati A, Scalenghe R (2021) Plastic end-of-life alternatives, with a focus on the agricultural sector. Curr Opin Chem Eng 32:100681. https://doi.org/10.1016/j.coche.2021.100681

20. Kader MA, Senge M, Mojid MA, Ito K (2017) Recent advances in mulching materials and methods for modifying soil environment. Soil Tillage Res 168:155–166. https://doi.org/10.1016/j.still.2017.01.001
21. Sintim HY, Flury M (2017) Is biodegradable plastic mulch the solution to agriculture's plastic problem? Environ Sci Technol 51:1068–1069. https://doi.org/10.1021/acs.est.6b06042
22. Ban D, Žanić K, Dumičić G, Čuljak TG, Ban SG (2009) The type of polyethylene mulch impacts vegetative growth, yield, and aphid populations in watermelon production. J Food Agric Environ 7:543–550
23. Ruiz JM, Hernandez J, Castilla N, Romero L (1999) Potato performance in response to different mulches. 1. Nitrogen metabolism and yield. J Agric Food Chem 47:2660–2665. https://doi.org/10.1021/jf981314x
24. Farias-Larios J, Orozco-Santos M (1997) Effect of polyethylene mulch colour on aphid populations, soil temperature, fruit quality, and yield of watermelon under tropical conditions. New Zeal J Crop Hortic Sci 25:369–374. https://doi.org/10.1080/01140671.1997.9514028
25. Zribi W, Aragüés R, Medina E, Faci JM (2015) Efficiency of inorganic and organic mulching materials for soil evaporation control. Soil Tillage Res 148:40–45. https://doi.org/10.1016/j.still.2014.12.003
26. Hughes BR, Zandstra J, Dale A (2012) Effects of mulch types on dayneutral strawberry production in three distinct environments in Ontario. Int J Fruit Sci 13:255–263. https://doi.org/10.1080/15538362.2013.713255
27. Torres-Olivar V, Ibarra-Jiménez L, Cárdenas-Flores A, Lira-Saldivar RH, Valenzuela-Soto JH, Castillo-Campohermoso MA (2017) Changes induced by plastic film mulches on soil temperature and their relevance in growth and fruit yield of pickling cucumber. Acta Agric Scand B Soil Plant Sci 68:97–103. https://doi.org/10.1080/09064710.2017.1367836
28. Ibarra-Jiménez L, Zermeño-González A, Munguía-López J, Quezada-Martín MAR, La Rosa-Ibarra MD (2008) Photosynthesis, soil temperature and yield of cucumber as affected by colored plastic mulch. Acta Agric Scand B Soil Plant Sci 58:372–378. https://doi.org/10.1080/09064710801920297
29. Zhang YL, Wang FX, Shock CC, Yang KJ, Kang SZ, Qin JT, Li SE (2017) Effects of plastic mulch on the radiative and thermal conditions and potato growth under drip irrigation in arid Northwest China. Soil Tillage Res 172:1–11. https://doi.org/10.1016/j.still.2017.04.010
30. APE EUROPE (2020) Plasticulture. https://apeeurope.eu/documentations-press-ape-europe/. Accessed 2 July 2021
31. Martín-Lara MA, Godoy V, Quesada L, Lozano EJ, Calero M (2021) Environmental status of marine plastics pollution in Spain. Mar Pollut Bull 170. https://doi.org/10.1016/j.marpolbul.2021.112677
32. Huang Y, Liu Q, Jia W, Yan C, Wang J (2020) Agricultural plastic mulching as a source of microplastics in the terrestrial environment. Environ Pollut 260:114096. https://doi.org/10.1016/j.envpol.2020.114096
33. Meng F, Fan T, Yang X, Riksen M, Xu M, Geissen V (2020) Effects of plastic mulching on the accumulation and distribution of macro and micro plastics in soils of two farming systems in Northwest China. PeerJ 8:e10375. https://doi.org/10.7717/peerj.10375
34. Zylstra ER (2013) Accumulation of wind-dispersed trash in desert environments. J Arid Environ 89:13–15. https://doi.org/10.1016/j.jaridenv.2012.10.004
35. Qi Y, Yang X, Pelaez AM, Huerta Lwanga E, Beriot N, Gertsen H, Garbeva P, Geissen V (2018) Macro- and micro-plastics in soil-plant system: effects of plastic mulch film residues on wheat (Triticum aestivum) growth. Sci Total Environ 645:1048–1056. https://doi.org/10.1016/j.scitotenv.2018.07.229
36. Beriot N, Zomer P, Zornoza R, Geissen V (2020) A laboratory comparison of the interactions between three plastic mulch types and 38 active substances found in pesticides. PeerJ 8:e9876. https://doi.org/10.7717/peerj.9876
37. Cicloplast (n.d.) Agricultural plastics industry adopts the European Plasticulture Strategy promoted by APE Europe. http://www.cicloplast.com/index.php?accion=notas-de-prensa&

subAccion=ver-noticia&id=95&page=1&frm%5bkeyword%5d=&actopc=42. Accessed 2 July 2021
38. EPRO (n.d.). https://www.epro-plasticsrecycling.org/. Accessed 2 July 2021
39. Guo JJ, Huang XP, Xiang L, Wang YZ, Li YW, Li H, Cai QY, Mo CH, Wong MH (2020) Source, migration and toxicology of microplastics in soil. Environ Int 137:105263. https://doi.org/10.1016/j.envint.2019.105263
40. Qi R, Jones DL, Li Z, Liu Q, Yan C (2020) Behaviour of microplastics and plastic film residues in the soil environment: a critical review. Sci Total Environ 703:134722. https://doi.org/10.1016/j.scitotenv.2019.134722
41. Yan L, Peng W (2021) Research of new pollutant microplastics in soil. IOP Conf Ser Earth Environ Sci 781:052005. https://doi.org/10.1088/1755-1315/781/5/052005
42. Fakour H, Lo SL, Yoashi NT, Massao AM, Lema NN, Mkhontfo FB, Jomalema PC, Jumanne NS, Mbuya BH, Mtweve JT, Imani M (2021) Quantification and analysis of microplastics in farmland soils: characterization, sources, and pathways. Agriculture 11:330. https://doi.org/10.3390/agriculture11040330
43. Yang L, Zhang Y, Kang S, Wang Z, Wu C (2021) Microplastics in soil: a review on methods, occurrence, sources, and potential risk. Sci Total Environ 780:146546. https://doi.org/10.1016/j.scitotenv.2021.146546
44. Piehl S, Leibner A, Löder MGJ, Dris R, Bogner C, Laforsch C (2018) Identification and quantification of macro- and microplastics on an agricultural farmland. Sci Rep 8:17950. https://doi.org/10.1038/s41598-018-36172-y
45. Li X, Chen L, Mei Q, Dong B, Dai X, Ding G, Zeng EY (2018) Microplastics in sewage sludge from the wastewater treatment plants in China. Water Res 142:75–85. https://doi.org/10.1016/j.watres.2018.05.034
46. Schothorst B, Beriot N, Lwanga EH, Geissen V (2021) Sources of light density microplastic related to two agricultural practices: the use of compost and plastic mulch. Environment 2021(8):36. https://doi.org/10.3390/environments8040036
47. Corradini F, Casado F, Leiva C, Huerta-Lwanga E, Geissen V (2021) Microplastics occurrence and frequency in soils under different land uses. Sci Total Environ 752:141917. https://doi.org/10.1016/j.scitotenv.2020.141917
48. Meixner K, Kubiczek M, Fritz I (2020) Microplastic in soil–current status in Europe with special focus on method tests with Austrian samples. AIMS Environ Sci 7:174–191. https://doi.org/10.3934/environsci.2020011
49. Weithmann N, Möller JN, Löder MGJ, Piehl S, Laforsch C, Freitag R (2018) Organic fertilizer as a vehicle for the entry of microplastic into the environment. Sci Adv 4:eaap8060. https://doi.org/10.1126/sciadv.aap8060
50. Zhang Y, Kang S, Allen S, Allen D, Gao T, Sillampää M (2020) Atmospheric microplastics: a review on current status and perspectives. Earth Sci Rev 203:103118. https://doi.org/10.1016/j.earscirev.2020.103118
51. Allen S, Allen D, Phoenix VR, Roux G, Durántez Jiménez P, Simomeau A, Binet S, Galop D (2019) Atmospheric transport and deposition of microplastics in a remote mountain catchment. Nat Geosci 12:339–344. https://doi.org/10.1038/s41561-019-0335-5
52. Liu K, Wang X, Fang T, Xu P, Zhu L, Li D (2019) Source and potential risk assessment of suspended atmospheric microplastics in Shanghai. Sci Total Environ 675:462–471. https://doi.org/10.1016/j.scitotenv.2019.04.110
53. Beriot N, Peek J, Zornoza R, Geissen V, Lwanga EH (2021) Low density-microplastics detected in sheep faeces and soil: a case study from the intensive vegetable farming in Southeast Spain. Sci Total Environ 755:142653. https://doi.org/10.1016/j.scitotenv.2020.142653
54. Huerta-Lwanga E, Mendoza-Vega J, Quej V, Chi J, Sanchez del Cid L, Chi C, Escalona-Segura G, Gertsen H, Salánki T, Van der Ploeg M, Koelmans AA, Geissen V (2017) Field evidence for transfer of plastic debris along a terrestrial food chain. Nat Sci Rep 7:14071. https://doi.org/10.1038/s41598-017-14588-2

55. Ljung E, Olesen K, Andersson P, Fältström E, Vollertsen J, Wittgren H, Hagman M (2018) Mikroplaster i kretsloppet. Svenskt Vatten Utveckling Program, Rapport N° 2018-13, 48 p
56. Rezaei M, Riksen M, Sirjani E, Sameni A, Geissen V (2019) Wind erosion as a driver for transport light density microplastics. Sci Total Environ 669:273–281. https://doi.org/10.1016/j.scitotenv.2019.02.382
57. Isari E, Papaioannou D, Kalavrouziotis I, Karapanagioti H (2021) Microplastics in agricultural soils: a case study in cultivation of watermelons and canning tomatoes. Water 13:2168. https://doi.org/10.3390/w13162168
58. Zhou B, Wang J, Zhang H, Shi H, Fei Y, Huang S, Tong Y, Wen D, Luo Y, Barceló D (2020) Microplastics in agricultural soils on the coastal plain of Hangzhou Bay, east China: multiple sources other than plastic mulching film. J Hazard Mater 388:121814. https://doi.org/10.1016/j.jhazmat.2019.121814
59. Betageri VM, Kottiswaran SV (2018) Effects of plastic mulching on yield of grafted brinjal. Int J Curr Microbial Appl Sci 7(7):2516–2522
60. de Souza Machado AA, Lau CW, Kloas W, Bergmann J, Bachelier JB, Faltin E, Becker R, Gorlich AS, Rillig MC (2019) Microplastics can change soil properties and affect plant performance. Environ Sci Technol 53:6044–6052
61. Duppong LM, Delate K, Liebman M, Horton R, Romerom F, Kraus G, Petrich J, Chowdbury PK (2006) Crop Sci 44(3):861–869
62. Kwon YS, Park SK, Ko KD (1988) Effect of different mulch materials on the soil environment, growth and yield of red pepper (*Capsicum annum* L.). Res Rep Rural Dev Adm Hort Korea Republic 30:9–17
63. Merline Sheela A, Shanmugasundaram R, Sundaram MD (2005) Effect of organic amendments on the oxygen uptake of Pseudomonas putida- PAPs -1 in chromium contaminated pond ash. Soil Sediment Contam 14(1):71–84
64. Singh PN, Joshi BP, Singh G (1998) Effect of mulch on moisture conservation, irrigation requirement and yield of potato. Indian J Agron 32(451):451
65. Steinmetz Z, Wollmann C, Schaefer M, Buchmann C, David J, Tröger J, Muñoz K, Frör O, Schaumann GE (2016) Plastic mulching in agriculture. Trading short-term agronomic benefits for long-term soil degradation? Sci Total Environ 550:690–705. https://doi.org/10.1016/j.scitotenv.2016.01.153
66. Liu EK, He WQ, Yan CR (2014) 'White revolution' to 'white pollution'–agricultural plastic film mulch in China. Environ Res Lett 9:091001. https://doi.org/10.1088/1748-9326/9/9/091001
67. Zhang D, Ng EL, Hu W, Wang H, Galaviz P, Yang H, Sun W, Li C, Ma X, Fu B, Zhao P, Zhang F, Jin S, Zhou M, Du L, Peng C, Zhang X, Xu Z, Xi B, Liu X, Sun S, Cheng Z, Jiang L, Wang Y, Gong L, Kou C, Li Y, Ma Y, Huang D, Zhu J, Yao J, Lin C, Qin S, Zhou L, He B, Chen D, Li H, Zhai L, Lei Q, Wu S, Zhang Y, Pan J, Gu B, Liu H (2020) Plastic pollution in croplands threatens long-term food security. Glob Chang Biol. https://doi.org/10.1111/gcb.15043
68. Smith M, Love DC, Rochman C, Neff RA (2018) Microplastic in seafood and implications of human health. Curr Environ Health Rep 5(3):375–386
69. Kumar V, Merline Sheela A (2021) Effect of plastic film mulching on the distribution of plastic residues in agricultural fields. Chemosphere 273:128590. https://doi.org/10.1016/j.chemosphere.2020.128590
70. Li W, Wufuer R, Duo J, Wang S, Luo Y, Zhang D, Pan X (2020) Microplastics in agricultural soils: extraction and characterization after different periods of polyethylene film mulching in an arid region. Sci Total Environ 749:141420
71. Zhu D, Chen QL, An XL, Yang XR, Christie P, Ke X, Wu LH, Zhu YG (2018) Exposure of soil collembolans to microplastics perturbs their gut microbiota and alters their isotopic composition. Soil Biol Biochem 116:302–310
72. Maab S, Daphi D, Lehmann A, Rilling MC (2017) Transport of microplastic by two collembolan species. Environ Pollut 225:456–459. https://doi.org/10.1016/j.envpol.2017.03.009

73. Gaylor MO, Harvey E, Hale RC (2013) Polybrominated diphenyl ether (PBDE) accumulation by earthworms (Eisenia fetida) exposed to biosolids-, polyurethane foam microparticle-, and Penta-BDE-amended soils. Environ Sci Technol 47:13831–13839. https://doi.org/10.1021/es403750a
74. Oliviero M, Tato T, Schiavo S, Fernández V, Manzo S, Beiras R (2019) Leachates of micronized plastic toys provoke embryotoxic effects upon sea urchin Paracentrotus lividus. Environ Pollut 247:706–715. https://doi.org/10.1016/j.envpol.2019.01.098
75. Wei W, Huang QS, Sun J, Wang JY, Wu SL, Ni BJ (2019) Polyvinyl chloride microplastics affect methane production from the anaerobic digestion of waste activated sludge through leaching toxic bisphenol-A. Environ Sci Technol 53(5):2509–2517. https://doi.org/10.1021/acs.est.8b07069
76. Jiang XJ, Liu W, Wang E, Zhou T, Xin P (2017) Residual plastic mulch fragments effects on soil physical properties and water flow behavior in the Minqin Oasis, northwestern China. Soil Tillage Res 166:100–107. https://doi.org/10.1016/j.still.2016.10.011
77. Bolan N, Kirkham MB, Halsband C, Nugegoda D, Ok YS (2020) Particulate plastics in terrestrial and aquatic environments. CRC Press. ISBN 9780367511401
78. Hahladakis JN, Velis CA, Weber R, Iacovidou E, Purnell P (2018) An overview of chemical additives present in plastics: migration, release, fate and environmental impact during their use, disposal and recycling. J Hazard Mater 344:179–199. https://doi.org/10.1016/j.jhazmat.2017.10.014
79. Kumar M, Xiong X, He M, Tsang DCW, Gupta J, Khan E, Harrad S, Hou D, Ok YS, Bolan NS (2020) Microplastics as pollutants in agricultural soils. Environ Pollut 265:114980. https://doi.org/10.1016/j.envpol.2020.114980
80. Fu X, Du Q (2011) Uptake of di-(2-ethylhexyl) phthalate of vegetables from plastic film greenhouses. J Agric Food Chem 59:11585–11588. https://doi.org/10.1021/jf203502e
81. Hu X, Bei W, Shan X (2003) Survey of phthalate pollution in arable soils in China. J Environ Monit 5:649. https://doi.org/10.1039/b304669a
82. Niu L, Xu Y, Xu C, Yun L, Liu W (2014) Status of phthalate esters contamination in agricultural soils across China and associated health risks. Environ Pollut 195:16–23. https://doi.org/10.1016/j.envpol.2014.08.014
83. Wang J, Luo Y, Teng Y, Ma W, Christie P, Li Z (2013) Soil contamination by phthalate esters in Chinese intensive vegetable production systems with different modes of use of plastic film. Environ Pollut 180:265–273. https://doi.org/10.1016/j.envpol.2013.05.036
84. Wang Y, Wang F, Xiang L, Gu C, Redmile-Gordon M, Sheng H, Wang Z, Fu Y, Bian Y, Jiang X (2021) Risk assessment of agricultural plastic films based on release kinetics of phthalate acid esters. Environ Sci Technol 55:3676–3685. https://doi.org/10.1021/acs.est.0c07008
85. He L, Gielen G, Bolan NS, Zhang X, Qin H, Huang H, Wang H (2015) Contamination and remediation of phthalic acid esters in agricultural soils in China: a review. Agron Sustain Dev 35:519–534. https://doi.org/10.1007/s13593-014-0270-1
86. Zeng F, Cui K, Xie Z, Wu L, Liu M, Sun G, Lin Y, Luo D, Zeng Z (2008) Phthalate esters (PAEs): emerging organic contaminants in agricultural soils in peri-urban areas around Guangzhou, China. Environ Pollut 156:425–434. https://doi.org/10.1016/j.envpol.2008.01.045
87. Du QZ, Fu XW, Xia HL (2009) Uptake of di-(2-ethylhexyl) phthalate from plastic mulch film by vegetable plants. Food Addit Contam Part A Chem Anal Control Expo Risk Assess 26:1325–1329. https://doi.org/10.1080/02652030903081952
88. Ma T, Teng Y, Christie P, Luo Y (2014) Phytotoxicity in seven higher plant species exposed to di-n-butyl phthalate or bis (2-ethylhexyl) phthalate. Front Environ Sci Eng 9:259–268. https://doi.org/10.1007/s11783-014-0652-2
89. Wang J, Lv S, Zhang M, Chen G, Zhu T, Zhang S, Teng Y, Christie P, Luo Y (2016) Effects of plastic film residues on occurrence of phthalates and microbial activity in soils. Chemosphere 151:171–177. https://doi.org/10.1016/j.chemosphere.2016.02.076

90. Wang D, Xi Y, Shi X-Y, Zhong Y-J, Guo C-L, Han Y-N, Li F-M (2021) Effect of plastic film mulching and film residues on phthalate esters concentrations in soil and plants, and its risk assessment. Environ Pollut 286:117546. https://doi.org/10.1016/j.envpol.2021.117546
91. Bandopadhyay S, Martin-Closas L, Pelacho A, DeBruyn JM (2018) Biodegradable plastic mulch films: impacts on soil microbial communities and ecosystem functions. Front Microbiol 9:819. https://doi.org/10.3389/fmicb.2018.00819
92. Ramos L, Berenstein G, Hughes EA, Zalts A, Montserrat JM (2015) Polyethylene film incorporation into the horticultural soil of small periurban production units in Argentina. Sci Total Environ 523:74–81. https://doi.org/10.1016/j.scitotenv.2015.03.142
93. Lwanga EH, Vega JM, Quej VK, Chi JA, del Cid LS, Chi C, Segura GE, Gertsen H, Salánki T, van der Ploeg M, Koelmans AA, Geissen V (2017) Field evidence for transfer of plastic debris along a terrestrial food chain. Sci Rep 7:14071. https://doi.org/10.1038/s41598-017-14588-2
94. Li Z, Li R, Li Q, Zhou J, Wang G (2020) Physiological response of cucumber (Cucumis sativus L.) leaves to polystyrene nanoplastics pollution. Chemosphere 255:127041. https://doi.org/10.1016/j.chemosphere.2020.127041
95. Lwanga EH, Gertsen H, Gooren H, Peters P, Salanki T, van der Ploeg M, Besseling E, Koelmans AA, Geissen V (2017) Incorporation of microplastics from litter into burrows of lumbricus terrestris. Environ Pollut 220:523–531
96. Iqbal S, Xu J, Allen SD, Khan S, Nadir S, Arif MS, Yasmeen T (2020) Unraveling consequences of soil micro- and nano-plastic pollution on soil-plant system: implications for nitrogen (N) cycling and soil microbial activity. Chemosphere 260:127578. https://doi.org/10.1016/j.chemosphere.2020.127578
97. SARE (2007) Managing cover crops profitability, 3rd edn. Handbook Series Book 9. Sustainable Agriculture Research and Education (SARE) program, with funding from the National Institute of Food and Agriculture, U.S. Department of Agriculture, 248 p
98. Tepe E, Hoover E, Poppe S (2008) The wool mulch system of producing strawberries. A manual for commercial growers in Minnesota. University of Minnesota (ed), 34 p
99. Tillman J, Nair A, Batzer J, Gleason M (2015) Strip-tillage and row cover use in organically and conventionally grown summer squash. Iowa State Res Farm Progress Rep 2187:7–9
100. Calero M, Godoy V, Quesada L, Martín-Lara MA (2021) Green strategies for microplastics reduction. Curr Opin Green Sustain Chem 28:100442. https://doi.org/10.1016/j.cogsc.2020.100442
101. Brodhagen M, Peyron M, Miles C, Inglis DA (2015) Biodegradable plastic agricultural mulches and key features of microbial degradation. Appl Microbiol Biotechnol 99:1039–1056. https://doi.org/10.1007/s00253-014-6267-5
102. Moreno MM, Moreno A (2008) Effect of different biodegradable and polyethylene mulches on soil properties and production in a tomato crop. Sci Hortic 116:256–263. https://doi.org/10.1016/j.scienta.2008.01.00
103. Touchaleaume F, Martin-Closas L, Angellier-Coussy H, Chevillard A, Cesar G, Gontard N, Gastaldi E (2016) Performance and environmental impact of biodegradable polymers as agricultural mulching films. Chemosphere 144:433–439. https://doi.org/10.1016/j.chemosphere.2015.09.006
104. Bandopadhyay S, Liquet Y González J, Henderson K, Anunciado M, Hayes D, De Bruyn J (2020) Soil microbial communities associated with biodegradable plastic mulch films. Front Microbiol 11:587074. https://doi.org/10.3389/fmicb.2020.587074
105. Beriot N (2020) EIPG-AGRI Focus Group. Reducing the plastic footprint of agriculture. Starting paper. May 2020. 29 p
106. Ardisson GB, Tosin M, Barbale M, Degli-Innocenti F (2014) Biodegradation of plastics in soil and effects on nitrification activity. A laboratory approach. Front Microbiol 5:710. https://doi.org/10.3389/fmicb.2014.00710
107. Sforzini S, Oliveri L, Chinaglia S, Viarengo A (2016) Application of biotests for the determination of soil ecotoxicity after exposure to biodegradable plastics. Front Environ Sci 4:68. https://doi.org/10.3389/fenvs.2016.00068

108. Koitabashi M, Noguchi MT, Sameshima-Yamashita Y, Hiradate S, Suzuki K, Yoshida S et al (2012) Degradation of biodegradable plastic mulch films in soil environment by phylloplane fungi isolated from gramineous plants. AMB Express 2:40. https://doi.org/10.1186/2191-0855-2-40
109. Muroi F, Tachibana Y, Kobayashi Y, Sakurai T, Kasuya K (2016) Influences of poly(butylene adipate-co-terephthalate) on soil microbiota and plant growth. Polym Degrad Stab 129:338–346. https://doi.org/10.1016/j.polymdegradstab.2016.05.018
110. Serrano-Ruiz H, Martin-Closas L, Pelacho A (2021) Biodegradable plastic mulches: impact on the agricultural biotic environment. Sci Total Environ 750:141228. https://doi.org/10.1016/j.scitotenv.2020.141228
111. Zumstein MT, Schintlmeister A, Nelson TF, Baumgartner R, Woebken D, Wagner M, Kohler HE, McNeill K, Sander M (2018) Biodegradation of synthetic polymers in soils: tracking carbon into CO_2 and microbial biomass. Sci Adv 4:eaas9024. https://doi.org/10.1126/sciadv.aas9024
112. Ahmed T, Shahid M, Azeem F, Rasul I, Shah A, Noman M, Hameed A, Manzoor N, Manzoor I, Muhammad S (2018) Biodegradation of plastics: current scenario and future prospects for environmental safety. Environ Sci Pollut Res 25:7287–7298. https://doi.org/10.1007/s11356-018-1234-9
113. Goldberger JR, Jones RE, Miles CA, Wallace RW, Inglis DA (2013) Barriers and bridges to the adoption of biodegradable plastic mulches for US specialty crop production. Renew Agric Food Syst 30(2):143–153. https://doi.org/10.1017/S1742170513000276
114. ASOBIOCOM (2018) Compostable and biodegradable plastic: an efficient and sustainable solution for agriculture report. 10 p
115. Shruti VC, Kutralam-Muniasamy G (2019) Bioplastics: missing link in the era of Microplastics. Sci Total Environ 697:134139. https://doi.org/10.1016/j.scitotenv.2019.134139
116. Green DS (2016) Effects of microplastics on European flat oysters, *Ostrea edulis* and their associated benthic communities. Environ Pollut 216:95–103. https://doi.org/10.1016/j.envpol.2016.05.043
117. Green DS, Boots B, Blockley DJ, Rocha C, Thompson R (2015) Impacts of discarded plastic bags on marine assemblages and ecosystem functioning. Environ Sci Technol 49:5380–5389
118. Zuo L-Z, Li H-X, Lin L, Sun Y-X, Diao Z-H, Liu S, Zhang Z-Y, Xu X-R (2019) Sorption and desorption of phenanthrene on biodegradable poly(butylene adipate coterephtalate) microplastics. Chemosphere 215:25–32
119. González-Pleiter M, Tamayo-Belda M, Pulido-Reyes G, Amariei G, Leganés F, Rosal R, Fernández-Piñas F (2019) Secondary nanoplastics released from a biodegradable microplastic severely impact freshwater environments. Environ Sci-Nano 6(5):1382–1392
120. Boots B, Russell CW, Green DS (2019) Effects of microplastics in soil ecosystems: above and below ground. Environ Sci Technol 53:11496–11506. https://doi.org/10.1021/acs.est.9b03304
121. Serrano-Ruiz H, Martin-Closas L, Pelacho AM (2018) Application of an in vitro plant ecotoxicity test to unused biodegradable mulches. Polym Degrad Stab 158:102–110. https://doi.org/10.1016/j.polymdegradstab.2018.10.016
122. Wang F, Zhang X, Zhang SS, Zhang SS, Sun Y (2020) Interactions of microplastics and cadmium on plant growth and arbuscular mycorrhizal fungal communities in an agricultural soil. Chemosphere 254:126791. https://doi.org/10.1016/j.chemosphere.2020.126791
123. HSMG (n.d.) Agricultural paper mulch, tag and ticket. https://www.hsmgrp.com/applications/agricultural-paper-mulch/. Accessed 28 July 2021
124. Haapala T, Palonen P, Korpela A, Ahokas J (2014) Feasibility of paper mulches in crop production: a review. Agric Food Sci 23:60–79. https://doi.org/10.23986/afsci.8542
125. Saglam M, Sintim HY, Bary AI, Miles CA, Ghimire S, Inglis DA, Flury M (2017) Modeling the effect of biodegradable paper and plastic mulch on soil moisture dynamics. Agric Water Manag 193:240–250. https://doi.org/10.1016/j.agwat.2017.08.011

126. Sánchez E, Lamont WJ, Orzolek MD (2008) Newspaper mulches for supressing weeds for organic high-tunnel cucumber production. Hort Technol 18:154–157
127. Ahokas J, Korpela A, Ince A, Güzel E, Asikainen J, Haapala T, Kujanpää M, Mikkola H, Pitkänen M, Tamminen A, Vikman M (2014) Paper based mulches as an alternative to polyethylene mulch in vegetable production. J Agric Machinery Sci 10(1):73–78
128. Barman KK (2016) Mulching as a component of organic/eco-friendly farming. In Conservation of natural resources and its efficient utilization for sustaining hill agriculture. ICAR Research Complex for NEG Region (ed), pp 26–30
129. Hoidal N (2021) Exploring alternatives to plastic mulch. https://blog-fruit-vegetable-ipm.extension.umn.edu/2021/01/exploring-alternatives-to-plastic-mulch.html. Accessed 28 July 2021
130. Peng Z, Ting W, Haixia W, Min W, Xiangping M, Siwei M, Rui Z, Zhikuan J, Qingfang H (2015) Effects of straw mulch on soil water and winter wheat production in dryland farming. Nat Sci Rep 5:10725. https://doi.org/10.1038/srep10725
131. Song D, Tariq A, Pan K, Chen W, Zhang A, Sun X, Ran Y, Zeng F (2020) Effects of straw mulching practices on soil nematode communities under walnut plantation. Nat Sci Rep 10: 15351. https://doi.org/10.1038/s41598-020-72530-5
132. Cook HF, Valdes GS, Lee HC (2006) Mulch effects on rainfall interception, soil physical characteristics and temperature under Zea mays L. Soil Tillage Res 91:227–235
133. Král M, Dvorák P, Capouchová I (2020) The effect of straw mulch and compost application on the soil lossed in potatoes cultivation. Plant Soil Environ 66(9):446–452. https://doi.org/10.17221/330/2020-PSE
134. Rowe B, Fernandez T, Cregg B (2004) Effect of wool mulch on propagation, crop growth, and weed control in liners. Propagation of Ornamental Plants 4(2):29–36
135. Kotodziejczyk M (2015) The effect of living mulches and conventional methods of weed control on weed infestation and potato yield. Sci Hortic 191:127–133. https://doi.org/10.1016/j.scienta.2015.05.016
136. Ginakes P, Grossman JM, Baker JM, Sooksa-nguan T (2020) Living mulch management spatially localizes nutrient cycling in organic corn production. Agriculture 10:243. https://doi.org/10.3390/agriculture10060243
137. James B, Trovati G, Peñalva C, Czech L, Mendioroz I, Milicic V (2021) EIP-AGRI Focus Group. Reducing the plastic footprint of agriculture. Minipaper D: Agricultural management, on site practice to reduce plastic use and the contamination in the environment. February 2021, 23 p
138. Marí A, Pardo G, Cirujeda A, Martínez Y (2019) Economic evaluation of biodegradable plastic films and paper mulches used in open-air grown pepper (*Capsicum annum* L.) crop. Agronomy 9:36. https://doi.org/10.3390/agronomy9010036

Removal Strategies for Aquatic Microplastics

Yingshuang Zhang, Hongru Jiang, Hui Wang, and Chongqing Wang

1 Removal Strategies for Aquatic Microplastics

1.1 Filtration Processing

Filtration is a separation processing to capture solid particles from discharging fluid based on the plastic size, assisted by a supporting media and vacuum system (Fig. 1a) [5]. Filtration includes membrane filtration and granular filtration. In membrane filtration, functional and selective barriers are available to retain specific molecules, ions, and particles. In granular filtration, granular media composed of glass beads, quartz sands, or activated carbon adsorb and preserve solid particles during attachment and transport [6].

Sequential filtration and screen set can hierarchically remove MPs with a wide size range, even nano-scale particles, benefiting from a cascade of retentates (Fig. 1b, c). Lv and Hernandez, respectively, analyzed the size of MPs in a municipal wastewater treatment plants (WWTPs) and personal care products by sequential filtration that might become an efficient removal processing for MPs [1, 2]. Wang et al. produced three biochar samples from corn straw and a hardwood and prepared granular biochar filters for the leaching column tests of MPs spheres. The filters provided above 95% removal capacity for the MPs spheres of 10 μm diameter, which was much higher than that of sand filters [7]. Tong et al. proposed porous media by adding biochar and Fe_3O_4-biochar in quartz sand to inhibit the transport of MPs particles. Approximate 16.5%, 14.6%, and 7.5% of 2 μm MPs were extracted by quartz sand, quartz sand with biochar amendment, and quartz sand with Fe_3O_4-

Y. Zhang · H. Jiang · H. Wang
College of Chemistry and Chemical Engineering, Central South University, Changsha, China
e-mail: hongru_jiang@qq.com

C. Wang (✉)
School of Chemical Engineering and Energy, Zhengzhou University, Zhengzhou City, China

A. Khan et al. (eds.), *Microplastic sources, fate and solution*,
https://doi.org/10.1007/978-981-99-0695-6_4

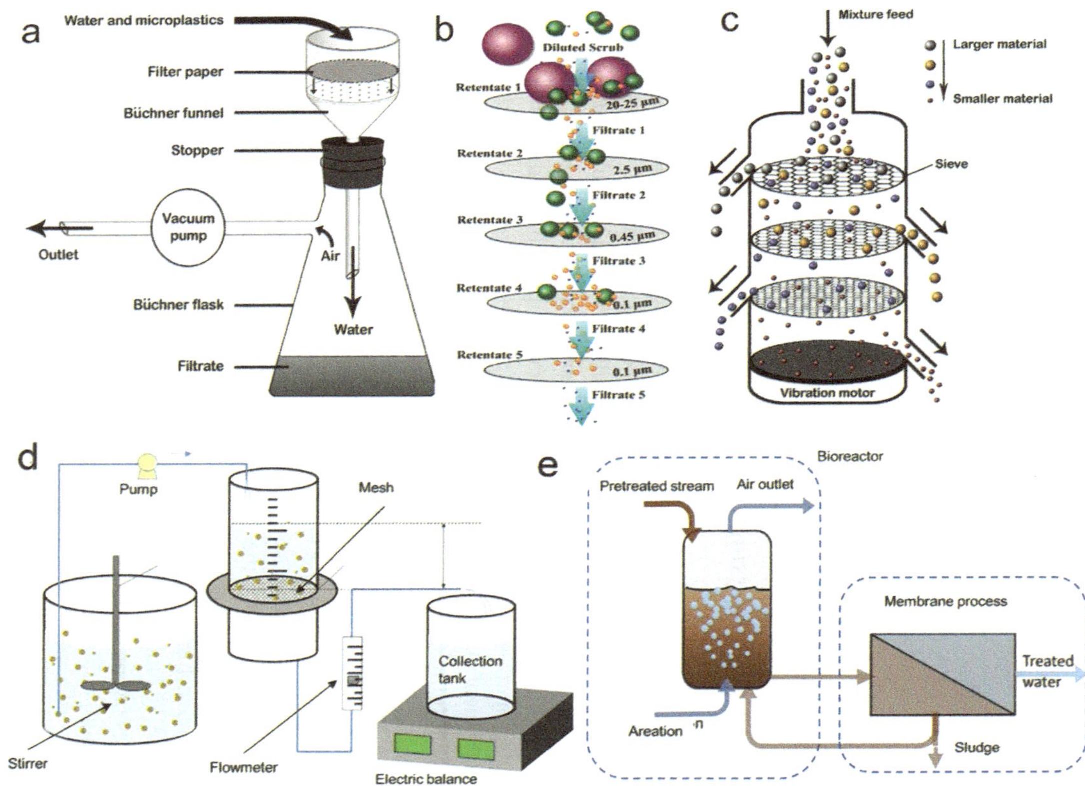

Fig. 1 Schematic diagram of filtration (**a**), sequential filtration (**b**), and multi-tier sieving (**c**), lab-scale external DM experimental setup (**d**), and membrane bioreactor process (**e**) [1–4]

biochar amendment [8]. Shen et al. developed aluminosilicate filter media and their surfactant-modified products to remove above 96% polyurethane and polyethylene MPs with particle size less than 10 μm [9].

The membrane filtration is divided into microfiltration, ultrafiltration, nanofiltration, and reverse osmosis filtration according to the pore size of membrane medium. Ma et al. applied ultrafiltration and Al/Fe-based coagulants into the polyethylene MPs removal in drinking water treatment. The ultrafiltration was an important section in polyethylene removal because ca. 80% MPs were discharged after coagulation even if flocs significantly increased the MPs particle size [10]. Besides, the presence of large PE particles ($2 < d < 5$ mm) only declined by 10% of membrane flux under 0.2 mM polyacrylamide (PAM) and 2 mM $FeCl_3{\cdot}6H_2O$ [11]. However, the membrane in this technique generally suffers from fouling due to improper chemical pretreatment and small MPs particles. Enfrin et al. suggested the positive effect of hydrophilic acrylic acid and cyclopropylamine plasma coatings on fouling mitigation and MPs filtration [12]. After crossflow filtration, the water flux of modified membrane declined less than 8% and the adsorption capccity MPs particles reduced by more than 60%.

Dynamic membrane filtration (DM) is a distinctive process using foulant as filtration cake layers that are secondary membranes when wastewater passes through supporting media [13, 14]. As shown in Fig. 1d, Li et al. operated DM technology on a 90 μm supporting mesh to remove 1.65–516 μm MPs from synthetic wastewater [3]. After 20 min of filtration, the effluent turbidity reduced to <1 NTU, total filtration resistance increased linearly with filtration time to 180 mm of water head, and transmembrane pressure increased to $6.52 \times 10^{-9}\ m^{-1}$. DM performed an excellent removal profile for low-density and non-degradable MPs, benefiting from the stable and light cake layers.

Membrane bioreactor (MBR), as a coupling technology of biological catalysts and membrane separation, involves pretreatment, bioreactor, and membrane filtration (Fig. 1e) [4, 15]. Talvitie et al. evaluated the contribution of MBR, disc filter, rapid sand filtration, and dissolved air flotation in WWTPs for MPs removal, of which removal ratio were 99.9%, 40–98.5%, 97%, and 95% [16]. Lares et al. also found a high MPs removal of 99.4% in MBR technology, where MPs in MBR permeate decreased ca. 60% compared to traditional activated sludge process [17]. The retention capacity for MPs of the studied WWTPs was 98.3%. Based on the research of Bayo et al., MBR and rapid sand filtration probed to statistically reduce 98.83% and 95.53% of microplastic particulate forms in WWTPs [18].

Based on the practicability, filtration processing was extensively applied in the effluent remediation of WWTPs and washing machines. Despite the efficient MPs removal, WWTPs could be a vital sink and source of MPs [19]. As shown in Fig. 2, WWTPs include preliminary, primary, secondary, and tertiary treatment processes. Course screening, fine screening, grit removal, and sedimentation can effectively remove 35–59% and 50–98% of MPs in preliminary and primary treatments [20, 21]. Filtration would be prior advanced final-stage treatment technologies in the tertiary treatment to remove the remaining 0.2–14% MPs after biological treatment and clarification in secondary treatment. Ziajahromi et al. investigated that

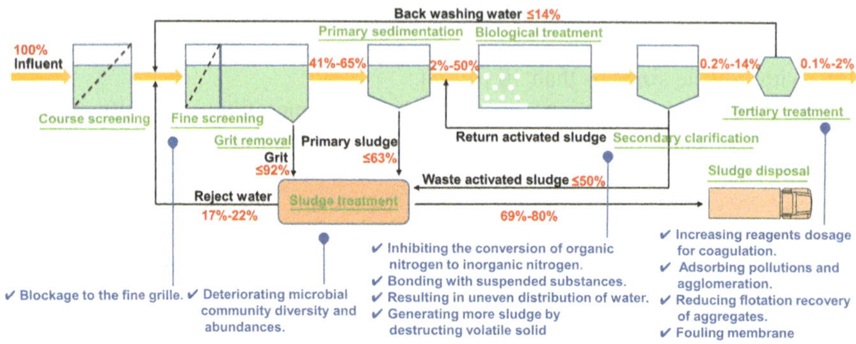

Fig. 2 Estimated MPs particle flow in wastewater treatment plant with primary, secondary, and tertiary treatment processes [20]

ultrafiltration and reverse osmosis could reduce the MPs concentration to 0.21 particles per liter at final stage [22]. Considering the annual effluxes of WWTPs, Mintenig et al. suggested post-filtration unit to further reduce the load of MPs and synthetic fibers substantially [23].

According to pertinent reports, synthetic fibers of textiles had increased to 62.9 Mt. in 2015, and upholstery, home furnishing, and clothing are significant contributors to microfibers during the regular washing [24, 25]. Mitigation strategies should be proposed to seize microfibers in the washing effluent, and the filtration became preferred separation processing. To reduce fiber releasing from clothes washing, McIlwraith et al. recommended the marketed Cora Ball and Lint LUV-R filters that decreased 26% and 87% of microfibers, respectively [26]. Plankton net, glass microfiber filters, and stainless-steel filter were also potentially applied to filtering microfibers from washing effluent [25]. Jedidi et al. reported a ceramic membrane composed of sintering supports and active layers from mineral coal fly ash to treat the textile dying effluents, of which turbidity, chemical oxygen demand values, and total color could be decreased significantly [27].

1.2 Density Separation

In density separation, a heavy medium is prepared to float the light MPs and separate MPs particles from sediments. An estimable and stable density of heavy medium is crucial for the density separation. The solutions of NaCl, NaBr, NaI, $ZnBr_2$, $CaCl_2$, $3Na_2WO_4{\cdot}9WO_3$, and NaH_2PO_4 were normally used as heavy media for the density separation of MPs, which can provide a wide density gradient in the range of 1.1–1.8 g cm^{-3} (Fig. 3a) [30–34].

Zhang et al. developed a novel density flotation to extract MPs (polyethylene, polypropylene, polystyrene, polymethyl methacrylate, polyethylene terephthalate,

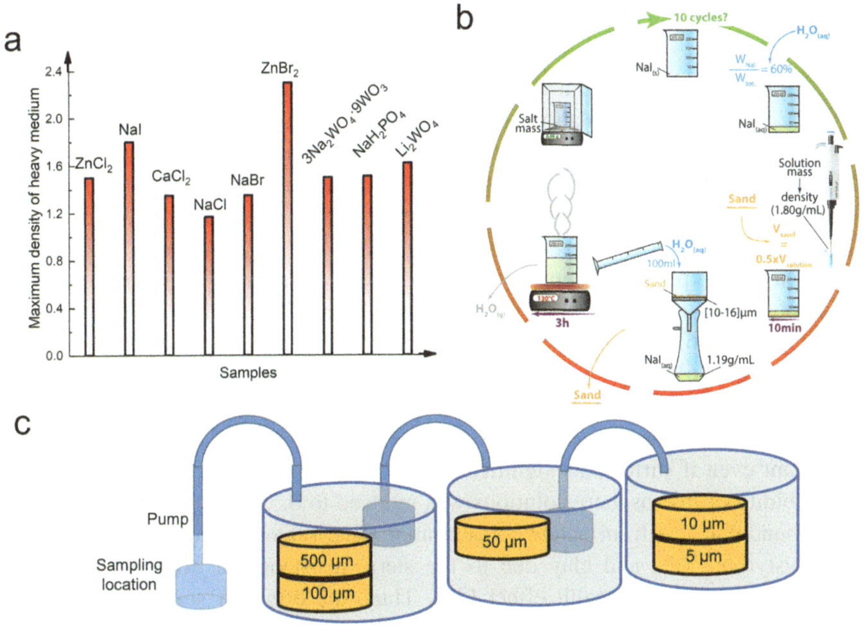

Fig. 3 The density of heavy media for density separation (**a**) and the schematic illustration of recyclable density flotation (**b**) [28] and fractionated filtration (**c**) [29]

polyoxymethylene, and polyvinyl chloride) from sediments with NaH_2PO_4 solution whose density changed in a wide density range of 1.4–1.5 g cm^{-3} by changing temperature in 20–40 °C [33]. Kedzierski et al. claimed that 1.80 g/mL NaI solution could extract 93–98% MPs produced in Europe with a total loss of 35.9% after the ten cycles of use, and NaI solutions are recyclable without any density alterations (Fig. 3b) [28]. Sánchez-Nieva et al. integrated NaCl saturated solution with aeration to extract and quantify the high-density polyethylene MPs, achieving 80% recovery for the low concentration of MPs and 90% recovery for high one [35]. Considering the size and density of MPs, Bannick et al. proposed a fractionated filtration system to separate and classify MPs particles from water, which not only prepared MPs samples but also provided the size distribution of MPs (Fig. 3c) [29]. Fuller et al. developed pressurized fluid extraction to segregate fine MPs particles generated through grinding and homogenizing and incorporated into the material [36]. The pressurized fluid could extract 101–111% of spiked plastics on glass beads and 85–94% on a composted municipal waste sample.

1.3 Coagulation and Agglomeration

Before physical separation, it is required to increase the particle size of suspended MPs by generating flocculation when sol becomes aggregate as unstable colloid [37]. Fe-based coagulants, Al-based coagulants, natural organic coagulants, and synthetic organic coagulants are commonly used as coagulants for MPs removal. Wang et al. declaimed a higher removal efficiency of granular activated carbon filtration (56.8–60.9%) than sedimentation (40.5–54.5%) combining coagulation [38]. Ma et al. explored the removal of polyethylene MPs that is the main constituent of microplastic in drinking water via coagulation and ultrafiltration. The removal ratios were 12.65%, 36.89%, 45.34%, and 61.19% under10 mM $FeCl_3•6H_2O$, 10 mM $AlCl_3•6H_2O$, 5 mM $AlCl_3•6H_2O$ + 15 mg/L cationic polyacrylamide (PAM), and 5 mM $AlCl_3•6H_2O$ + 15 mg/L anionic PAM (Fig. 4a) [10]. Skaf et al. also observed the formation of dense clusters with the help of PAM and Al-based coagulant even if surfactants significantly stabilized microsphere suspensions, and the turbidity of microsphere solutions was reduced to 0.5 NTU [43]. However, Xia et al. found that a high surfactant concentration deteriorated the coagulation removal of polystyrene MPs and clay due to the steric resistance of nonionic surfactant induced by surfactant stealth effect [44]. Thus, the effect of surfactants on MPs coagulation should be researched further in the future.

Zhang et al. conducted MPs coagulation removal combining with sedimentation and granular filtration. The highest removal was only 13.6% with coagulation and sedimentation, but 99.9% of removal capacity was obtained with coagulation and granular filtration [45]. Interestingly, biofilm increased the removal efficiency of sedimentation from <2.0% to 16.5%. Besides, MPs particles with small size (10–20 μm) and spherical-smooth shape could easily escape during coagulation. Similar results were also reported by Shahi et al. in researches about coagulation enhancement by using polyamine-coated sand [46]. Lapointe et al. proved that weathering improved the MPs affinity for Al-coagulants by changing the surface chemistry and roughness of MPs with 99% and 84% removal for fibers and microspheres under 0.3 mg/L cationic PAM and 2.73 mg/L Al [47].

Herbort et al. presented a pH-induced sol-gel process and sand trap separation to completely remove the polyethylene and polypropylene MPs from freshwater system (Fig. 4b). The n-alkyl-substituted chlorosilanes became a macromolecular hybrid silica network via the condensation of unstable silanol groups to include MPs particles [39]. The MPs trapping with short-chain (C1–C2) and long-chain (>C8) trichlorosilanes was inferior to that with medium molecular chain (C3–C5) counterparts. Further, Herbort et al. proposed alkoxy-silyl-induced co-agglomeration of hybrid silica gels and polymer particles to agglomerate, include, and remove MPs [40]. The trialkoxysilyl functionalized triazacyclohexane was determined as the best bioinspired molecules for MPs removal (Fig. 4c). Differently, Sturm et al. focused on alkyl structure of alkyltrichlorosilanes. The authors determined that linear precursors with C3–C5 were suitable for MPs removal, but molecules with branched

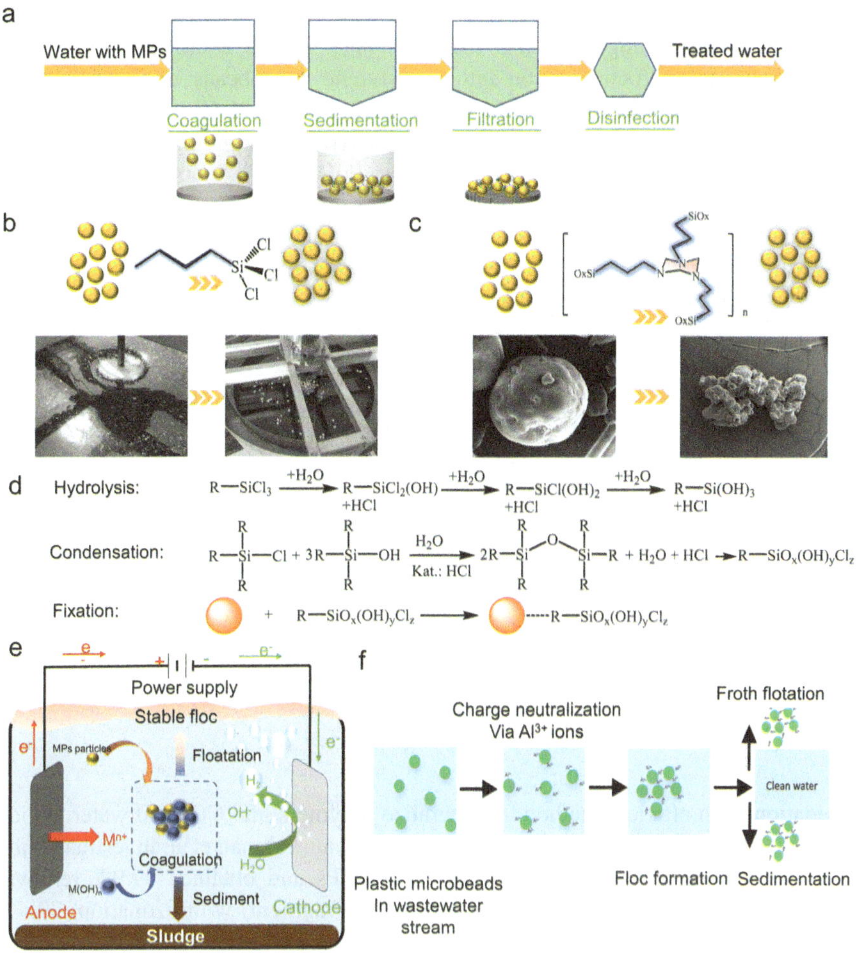

Fig. 4 (**a**) MPs removal process through coagulation and filtration, (**b**) MPs coagulation by trichlorosilanes and (**c**) trialkoxysilyl functionalized triazacyclohexane, (**d**) Hydrolysis, condensation, and fixation of alkyltrichlorosilanes for MPs removal [39, 40], (**e**) bench-scale electrocoagulation reactor [41], (**f**) MPs electrocoagulation for wasteater [42]

alkyl groups were unadvisable for MPs aggregation and removal due to the slow hydrolysis and condensation (Fig. 4d) [48].

1.4 Electrocoagulation

The electrocoagulation process includes cathodic reactions (Eqs. 1–2), anodic reactions (Eqs. 3–4), and coagulants generation, such as hydroxides of Fe^{2+} and Al^{3+}

(Eq. 5) [49]. Then, the coagulants trap the suspended MPs particles to form microflocs by forming sludge blankets (Fig. 4e) [41]. Perren et al. investigated the application of electrocoagulation into polystyrene microbeads removal, and the removal efficiency was slightly affected by the initial pH, NaCl concentration, and current density (Fig. 4f) [42]. As observed in typical experiments, the removal efficiencies of microbeads were more than 90% [50]. Akarsu et al. studied an electrocoagulation/electroflotation process for the treatment of laundry wastewater and optimized electrode type, initial pH, current, and application time. The removal efficiency for MPs particles was above 98% under Fe–Al electrode, pH 9, 2.16 A of current, and 60 min of reaction time [51].

$$M^{n+}_{(aq)} + ne^{-} \rightarrow M_{(s)} \tag{1}$$

$$2H_2O_l + 2e^{-} \rightarrow 2OH^{-} + H_{2(g)} \tag{2}$$

$$M_{(s)} \rightarrow M^{n+}_{(aq)} + ne^{-} \tag{3}$$

$$2H_2O_l \rightarrow 4H^{+}_{(aq)} + O_{2(g)} + 4e^{-} \tag{4}$$

$$M^{n+}_{(aq)} + nOH^{-} \rightarrow M(OH)_{n(s)} \tag{5}$$

1.5 Oxidation Process

Oxidation is an efficient method to mineralize polymers into CO_2 and water, which reduces the MPs content in wastewater [52]. Hidayaturrahman et al. investigated the removal of MPs from different stages of WWTPs and obtained 89.9% removal efficiency for MPs (microbead, fiber, sheet, and fragment) with ozonation, 79.4% with membrane disc filter, and 73.8% with rapid sand filtration [53]. Besides, Ahmed et al. reviewed the reliability of hybrid oxidation systems, such as ozonation/H_2O_2 and photo-Fenton, for chemical removal of emerging MPs contaminants [54]. This result was also supported by the study of Wang et al. A combination of ozonation and granular active carbon filtration increased 17.2–22.2% of removal efficiency for polystyrene MPs. However, the abundance of 1–5 μm MPs from ozonation effluent increased by 2.8–16.0%, resulting in a negative removal effect [38].

Tagg et al. attempted to break down organic compounds and isolate MPs particles from wastewater, but the chemistry or size changes of polymer particles were not observed [55]. Therefore, a modified Fenton process is required to remove MPs in water. Inspired by the superior features of transition metals and carbon catalysts, Kang et al. fabricated novel nanocomposites (Mn@NCNTs) by encapsulating manganese carbide nanoparticles in N-doped carbon nanotubes via one-pot pyrolysis and activated peroxymonosulfate to degrade cosmetic MPs particles under hydrothermal conditions (50% of removal efficiency) [56]. Electrochemical advanced oxidation was newly developed to degrade MPs debris by direct and mediate oxidation under in situ generated oxidant electrochemically [57]. Miao et al. prepared TiO_2/graphite cathode to degrade polyvinyl chloride MPs and remove ca. 75% particles based on cathodic reduction dechlorination and hydroxyl radical oxidation [58].

Liang et al. embedded photocatalytic polyacrylamide grafted TiO_2 (PAM-g-TiO_2) nanocomposite into low-density polyethylene and reduced its weight by 39.85% and average molecular weight by 94.60% [59]. Thus, photocatalytic oxidation is proposed as an attractive alternative for sustainable removal of MPs. Ariza-Tarazona et al. decreased 6.40% of high-density polyethylene MPs within 18 h, under visible light radiations in the presence of N-TiO_2-base semiconductors [60]. Tofa et al. photocatalytically degrade LDPE residues with visible light and zinc oxide nanorods [61]. Hydroxyl and superoxide radicals triggered chain breaking, branching, and crosslinking, producing low-molecular-weight polyethylene alkyl radicals. However, the mass loss of MPs debris was not reported in this study. Domínguez-Jaimes et al. degraded 23.5% of nano-polystyrene particles by photocatalysis using anodized mixed-structure TiO_2 under 30 °C and 50 h [62].

1.6 Magnetic Separation

After preparing magnetic seeds, the magnetism of MPs is enhanced by grafting magnetic seeds to polymer surfaces, and the magnetic MPs particles are extracted from wastewater by using magnetic fields. In this context, Grbic et al. created hydrophobic iron nanoparticles modified by hexadecyltrimethoxysilane, magnetized MPs particles by novel magnetic seeds, and recovered magnetic MPs particles from various environmental samples, such as seawater, freshwater, and sediments (Fig. 5a) [63]. Authors recovered 92% of 10–20 μm polyethylene and polystyrene beads, 93% of >1 mm polyethylene, polyethylene terephthalate, polystyrene, polyurethane, polyvinyl chloride, and polypropylene, and 78% of 0.2–1 mm polyethylene, polystyrene, polyurethane, polyvinyl chloride, and polypropylene. Tang et al. synthesized magnetic carbon nanotubes (M-CNTs) that was an effective adsorbent for polyethylene, polyethylene terephthalate, and polyamide MPs due to hydrophobic interaction, π-π electron conjugation, complexation, electrostatic interaction, and hydrogen–bond interaction. With the application of magnetic forces, the agglomerates of MPs/M−CNTs were completely removed within 300 min [65]. Rhein et al. isolated the MPs from aquatic systems by hetero-agglomeration and magnetic separation with the help of magnetic seeded filtration (Fig. 5b) [64]. The separation efficiency of MPs was 95% from dilute suspension.

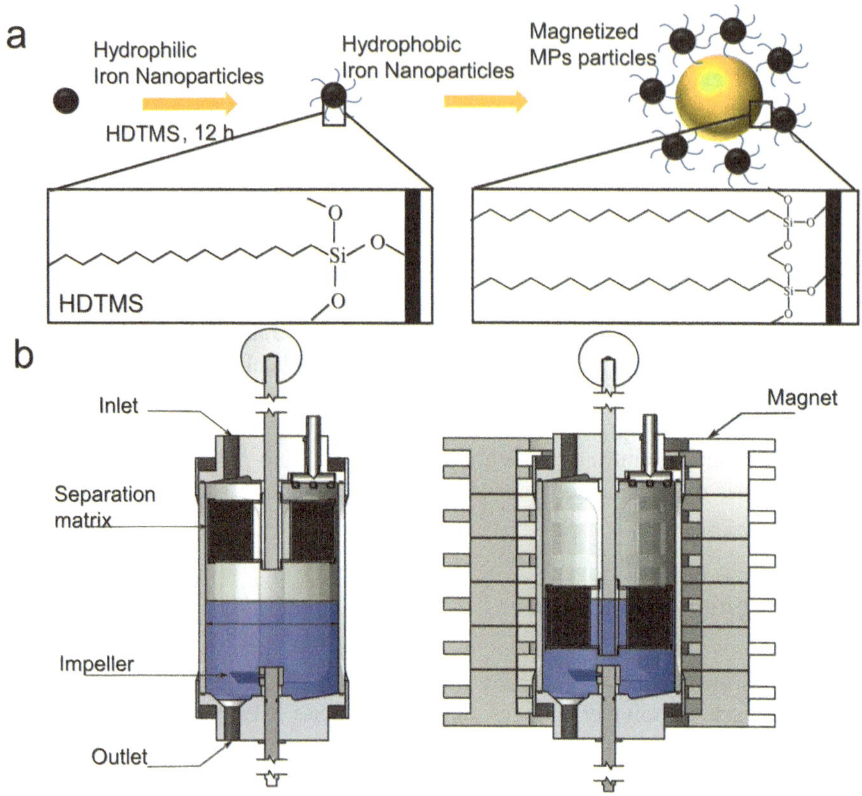

Fig. 5 Schematic of magnetic separation of microplastics by hydrophobic Fe nanoparticles modified by hexadecyltrimethoxysilane (**a**) [63]. Magnetic separation device for hetero-agglomeration and magnetic separation of microplastics (**b**) [64]

1.7 Oil Film Protocol

MPs separation using oil film is based on the hydrophobicity of MPs particles, which can take advantage of oleophilic properties of plastic. Mani et al. used castor oil to separate 0.3–1 mm polypropylene, polystyrene, polymethylmethacrylate, and polymethyl terephthalate MPs from suspended surface solids, soil, and sediment. The mean spike-recovery ratio of MPs was ca. 99% ± 4, and this separation protocol recovered 74% ± 13 of MPs from non-spiked environmental Rhine River suspended solids samples [66]. Crichton et al. presented an oil extraction protocol to separate MPs from aquatic sediments and observed 96.1% ± 7.4 recovery rates for total MPs, 92.7% ± 4.3 for fibers, and 99% ± 1.4 for particles with canola oil [30]. However, the small underflow nozzle of separatory funnel limited the upper size of recovered MPs [67]. Wang et al. developed a combination technology of refrigeration (−40 °C) and oil film separation to concentrate over 90% of MPs in olive oil by using

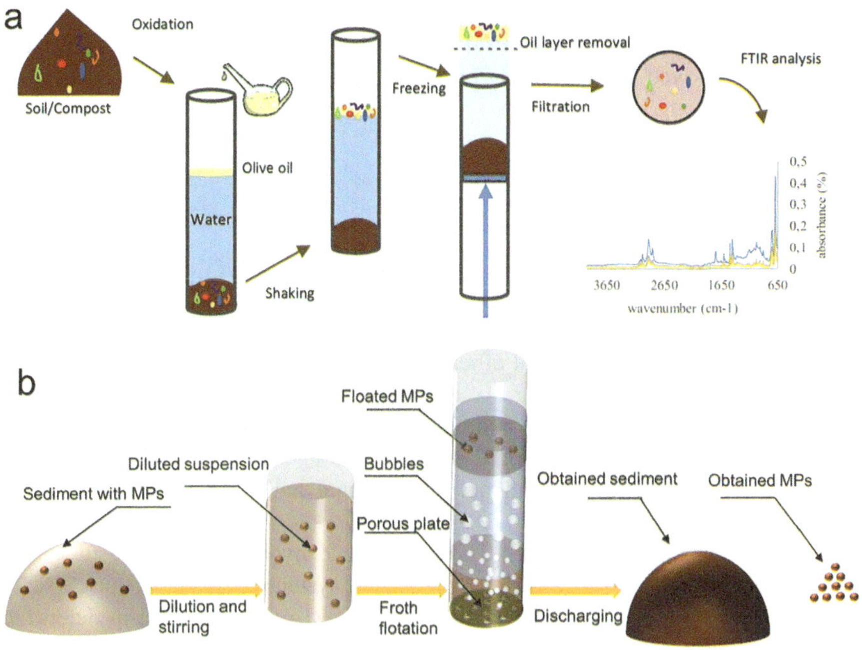

Fig. 6 Schematic illustration of oil film separation (**a**) and froth flotation (**b**) [68]

polytetrafluoroethylene cylinders (Fig. 6a) [68]. This method successfully avoided the limitation of separatory funnel.

1.8 Froth Flotation

Froth flotation becomes an efficient separation method for mineral processing based on a selective adherence of bubbles on hydrophobic surface. Post-consumer plastic generally performs a hydrophobic low-energy surface and is prone to be entrained with bubbles (Fig. 6b) [69]. By dispersing air bubbles into an environmental matrix, froth flotation can be a sample preparation method to collect MPs from the aquatic system [5]. Talvitie et al. designed a dissolved air flotation at the final stage of WWTPs to decrease 95% of discharging MPs [16]. Enfrin et al. and Sun et al. also considered froth flotation to improve the performance of WWTPs for MPs removal [20, 70]. Zhang et al. designed coagulative colloidal gas aphrons to remove polymethylmethacrylate and polystyrene MPs from aquatic systems. The functionalized bubbles by poly aluminum chloride and cetyltrimethylammonium bromide removed 94% MPs due to the synergistic effect of coagulation and flotation [71]. Zhang et al. researched the effect of size, type, metal ions, temperature, pH, aeration, frother dosage, and humic acid on flotation removal of polyethylene terephthalate and polystyrene MPs from artificial samples, and the MPs were completely removed under 30 mg/L humic acid, 0.05 mM Al^{3+}, 5.4 mL/min aeration, and 28 mg/L frother [72].

1.9 Adsorption Removal

Wang et al. fabricated natural and biodegradable sponge materials from plant protein through chemical crosslinking. The protein sponge with amino acid residues on side chains performed 81.2% removal efficiency by adsorbing MPs under pH 6–9 and initial MPs concentration of 1 mg L^{-1} (Fig. 7a) [73]. Sun et al. prepared robust and compressive chitin and graphene oxide sponges and conducted MPs adsorption from water at pH 6–8. The sponge could remove 89.8%, 72.4%, and 88.9% for neat polystyrene, carboxylate-modified polystyrene, and amine modified polystyrene, even after three cycles [75].

Tiwari et al. synthesized Zn-Al layered double hydroxides (LDH) to isolate hazardous nano-polystyrene particles from aquatic systems. The adsorption capacity of LDH for plastic debris were 164.49 mg/g, 162.62 mg/g, 53 mg/g in deionized water, synthetic freshwater, and hard water (Fig. 7b) [74]. Competitive binding and destabilization limited the removal efficiency of LDH for MPs. Chen et al. observed in situ phase transition of Mg/Al-flocs to Mg/Al-LDH in alkaline solution, and 90% of polystyrene nanoparticles were removed from water [76].

Singh et al. assessed the performance of iron-modified biochar pyrolyzed at 550 °C and 850 °C for MPs adsorption and extracted MPs from water under magnetic field. The maximum removal capacity was 225.11 mg/g for 1000 nm carboxyl polystyrene, 206.46 mg/g for 1000 nm amine-functionalized polystyrene, and 290.20 mg/g for 30 nm carboxyl-functionalized polystyrene [77]. Magid et al. enhanced adsorption of nano-polystyrene particles by using oxidized corncob

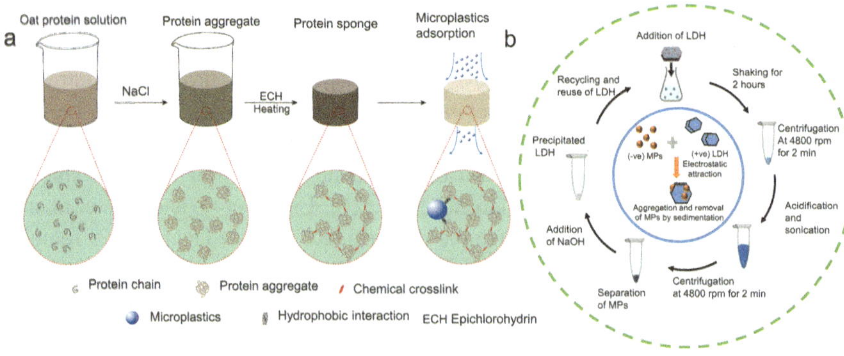

Fig. 7 (**a**) Biodegradable sponge materials and (**b**) Zn-Al layered double hydroxides for MPs adsorption [73, 74]

biochar pyrolyzed at 500 °C, 700 °C, and 900 °C, which obtained 18 mg/g adsorption capacity [78]. Ganie et al. converted agricultural waste to biochar and evaluated for their potential in removing nanoplastic particles. Biochar prepared at 750 °C, 550 °C, and 350 °C showed MPs removal efficiency of >99%, <39%, and < 24%, respectively [79].

1.10 Bioremediation

The bioremediation includes activated sludge process, biofilm process, biodegradation, constructed wetlands, and digestion. Lares et al. investigated the removal efficiency of MPs from WWTPs based on conventional activated sludge process, revealing a 98.3% removal for MPs, 89.8% removal for particles, and 99.1% removal for fibers [17]. To improve the removal efficiency of conventional sludge treatments, Chen et al. proposed hyperthermophilic composting technology to enhance the in situ biodegradation of MPs in sewage sludge. 43.7% of MPs were biodegraded from the sewage sludge after 45 days of hyperthermophilic composting treatment, which is the highest value ever reported for MPs biodegradation [80]. Lange et al. designed a vegetated bioretention cell to remove 70% of rubber, bitumen, and other MPs with size fractions of 100–300 μm and >300 μm from stormwater. To the contrast, gross pollutant trap was invalid for the removal of MPs with 100–300 μm in size [81]. Summers et al. explained that exopolymeric substances can lead to the agglomeration of nano- or micro-plastic particles [82]. Intriguingly, Cunha et al. applied extracellular polymeric substances from Cyanothece sp. strain, as a type of bioflocculant, into inducing MPs (<300 μm) coagulation. The authors thought that microalgal-based extracellular polymers could be a promising candidate to replace hazardous bioflocculants used in WWTPs [83].

References

1. Lv X, Dong Q, Zuo Z, Liu Y, Huang X, Wu W (2019) Microplastics in a municipal wastewater treatment plant: fate, dynamic distribution, removal efficiencies, and control strategies. J Clean Prod 225:579–586. https://doi.org/10.1016/j.jclepro.2019.03.321
2. Hernandez LM, Yousefi N, Tufenkji N (2017) Are there nanoplastics in your personal care products? Environ Sci Tech Let 4:280–285. https://doi.org/10.1021/acs.estlett.7b00187
3. Li L, Xu G, Yu H, Xing J (2018b) Dynamic membrane for micro-particle removal in wastewater treatment: performance and influencing factors. Sci Total Environ 627:332–340. https://doi.org/10.1016/j.scitotenv.2018.01.239
4. Poerio T, Piacentini E, Mazzei R (2019) Membrane processes for microplastic removal. Molecules 24:4148. https://doi.org/10.3390/molecules24224148
5. Crawford CB, Quinn B (2017) Microplastic separation techniques. In: Microplastic pollutants. Elsevier, Amsterdam, pp 203–218

6. Wu N, Wyart Y, Liu Y, Rose J, Moulin P (2013) An overview of solid/liquid separation methods and size fractionation techniques for engineered nanomaterials in aquatic environment. Environ Technol Rev 2:55–70. https://doi.org/10.1080/09593330.2013.788073
7. Wang ZH, Sedighi M, Lea-Langton A (2020a) Filtration of microplastic spheres by biochar: removal efficiency and immobilisation mechanisms. Water Res 184:116165. https://doi.org/10.1016/j.watres.2020.116165
8. Tong MP, He L, Rong HF, Li M, Kim HJ (2020) Transport behaviors of plastic particles in saturated quartz sand without and with biochar/Fe_3O_4-biochar amendment. Water Res 169:115284. https://doi.org/10.1016/j.watres.2019.115284
9. Shen MC, Hu T, Huang W, Song B, Zeng GM, Zhang YX (2021) Removal of microplastics from wastewater with aluminosilicate filter media and their surfactant-modified products: performance, mechanism and utilization. Chem Eng J 421:129918. https://doi.org/10.1016/j.cej.2021.129918
10. Ma B, Xue W, Hu C, Liu H, Qu J, Li L (2019a) Characteristics of microplastic removal via coagulation and ultrafiltration during drinking water treatment. Chem Eng J 359:159–167. https://doi.org/10.1016/j.cej.2018.11.155
11. Ma B, Xue W, Ding Y, Hu C, Liu H, Qu J (2019b) Removal characteristics of microplastics by Fe-based coagulants during drinking water treatment. J Environ Sci (China) 78:267–275. https://doi.org/10.1016/j.jes.2018.10.006
12. Enfrin M, Wang JS, Merenda A, Dumée LF, Lee J (2021) Mitigation of membrane fouling by nano/microplastics via surface chemistry control. J Membrane Sci 633:119379. https://doi.org/10.1016/j.memsci.2021.119379
13. Ma J, Wang Z, Xu Y, Wang Q, Wu Z, Grasmick A (2013) Organic matter recovery from municipal wastewater by using dynamic membrane separation process. Chem Eng J 219:190–199. https://doi.org/10.1016/j.cej.2012.12.085
14. Li L, Xu G, Yu H (2018a) Dynamic membrane filtration: formation, filtration, cleaning, and applications. Chem Eng Technol 41:7–18. https://doi.org/10.1002/ceat.201700095
15. Xiao K, Liang S, Wang X, Chen C, Huang X (2019) Current state and challenges of full-scale membrane bioreactor applications: a critical review. Bioresour Technol 271:473–481. https://doi.org/10.1016/j.biortech.2018.09.061
16. Talvitie J, Mikola A, Koistinen A, Setälä O (2017a) Solutions to microplastic pollution–removal of microplastics from wastewater effluent with advanced wastewater treatment technologies. Water Res 123:401–407. https://doi.org/10.1016/j.watres.2017.07.005
17. Lares M, Ncibi MC, Sillanpää M, Sillanpää M (2018) Occurrence, identification and removal of microplastic particles and fibers in conventional activated sludge process and advanced MBR technology. Water Res 133:236–246. https://doi.org/10.1016/j.watres.2018.01.049
18. Bayo J, López-Castellanos J, Olmos S (2020) Membrane bioreactor and rapid sand filtration for the removal of microplastics in an urban wastewater treatment plant. Mar Pollut Bull 156:111211. https://doi.org/10.1016/j.marpolbul.2020.111211
19. Murphy F, Ewins C, Carbonnier F, Quinn B (2016) Wastewater treatment works (WwTW) as a source of microplastics in the aquatic environment. Environ Sci Technol 50:5800–5808. https://doi.org/10.1021/acs.est.5b05416
20. Sun J, Dai X, Wang Q, van Loosdrecht MCM, Ni B (2019) Microplastics in wastewater treatment plants: detection, occurrence and removal. Water Res 152:21–37. https://doi.org/10.1016/j.watres.2018.12.050
21. Talvitie J, Mikola A, Setälä O, Heinonen M, Koistinen A (2017b) How well is microlitter purified from wastewater?–a detailed study on the stepwise removal of microlitter in a tertiary level wastewater treatment plant. Water Res 109:164–172. https://doi.org/10.1016/j.watres.2016.11.046
22. Ziajahromi S, Neale PA, Rintoul L, Leusch FDL (2017) Wastewater treatment plants as a pathway for microplastics: development of a new approach to sample wastewater-based microplastics. Water Res 112:93–99. https://doi.org/10.1016/j.watres.2017.01.042

23. Mintenig SM, Int-Veen I, Löder MGJ, Primpke S, Gerdts G (2017) Identification of microplastic in effluents of waste water treatment plants using focal plane array-based micro-Fourier-transform infrared imaging. Water Res 108:365–372. https://doi.org/10.1016/j.watres.2016.11.015
24. Hartline NL, Bruce NJ, Karba SN, Ruff EO, Sonar SU, Holden PA (2016) Microfiber masses recovered from conventional machine washing of new or aged garments. Environ Sci Technol 50:11532–11538. https://doi.org/10.1021/acs.est.6b03045
25. Schöpel B, Stamminger R (2019) A comprehensive literature study on microfibres from washing machines. Tenside Surfact Det 2:95–104. https://doi.org/10.3139/113.110610
26. McIlwraith HK, Lin J, Erdle LM, Mallos N, Diamond ML, Rochman CM (2019) Capturing microfibers–marketed technologies reduce microfiber emissions from washing machines. Mar Pollut Bull 139:40–45. https://doi.org/10.1016/j.marpolbul.2018.12.012
27. Jedidi I, Khemakhem S, Saïdi S, Larbot A, Elloumi-Ammar N, Fourati A, Charfi A, Salah AB, Amar RB (2011) Preparation of a new ceramic microfiltration membrane from mineral coal fly ash: application to the treatment of the textile dying effluents. Powder Technol 208:427–432. https://doi.org/10.1016/j.powtec.2010.08.039
28. Kedzierski M, Le Tilly V, César G, Sire O, Bruzaud S (2017) Efficient microplastics extraction from sand. A cost effective methodology based on sodium iodide recycling. Mar Pollut Bull 115:120–129. https://doi.org/10.1016/j.marpolbul.2016.12.002
29. Bannick CG, Szewzyk R, Ricking M, Schniegler S, Obermaier N, Barthel AK, Altmann K, Eisentraut P, Braun U (2019) Development and testing of a fractionated filtration for sampling of microplastics in water. Water Res 149:650–658
30. Crichton EM, Noel M, Gies EA, Ross PS (2017) A novel, density-independent and FTIR-compatible approach for the rapid extraction of microplastics from aquatic sediments. Anal Methods 9:1419–1428
31. Nuelle M, Dekiff JH, Remy D, Fries E (2014) A new analytical approach for monitoring microplastics in marine sediments. Environ Pollut 184:161–169. https://doi.org/10.1016/j.envpol.2013.07.027
32. Quinn B, Murphy F, Ewins C (2017) Validation of density separation for the rapid recovery of microplastics from sediment. Anal Methods-UK 9:1491–1498. https://doi.org/10.1039/c6ay02542k
33. Zhang X, Yu K, Zhang H, Liu Y, He J, Liu X, Jiang J (2020a) A novel heating-assisted density separation method for extracting microplastics from sediments. Chemosphere 256:127039. https://doi.org/10.1016/j.chemosphere.2020.127039
34. Li Q, Wu J, Zhao X, Gu X, Ji R (2019) Separation and identification of microplastics from soil and sewage sludge. Environ Pollut 254:113076. https://doi.org/10.1016/j.envpol.2019.113076
35. Sánchez-Nieva J, Perales JA, González-Leal JM, Rojo-Nieto E (2017) A new analytical technique for the extraction and quantification of microplastics in marine sediments focused on easy implementation and repeatability. Anal Methods-UK 9:6371–6378. https://doi.org/10.1039/C7AY01800B
36. Fuller S, Gautam A (2016) A procedure for measuring microplastics using pressurized fluid extraction. Environ Sci Technol 50:5774–5780. https://doi.org/10.1021/acs.est.6b00816
37. Nakazawa Y, Matsui Y, Hanamura Y, Shinno K, Shirasaki N, Matsushita T (2018) Identifying, counting, and characterizing superfine activated-carbon particles remaining after coagulation, sedimentation, and sand filtration. Water Res 138:160–168. https://doi.org/10.1016/j.watres.2018.03.046
38. Wang Z, Lin T, Chen W (2020b) Occurrence and removal of microplastics in an advanced drinking water treatment plant (ADWTP). Sci Total Environ 700:134520. https://doi.org/10.1016/j.scitotenv.2019.134520
39. Herbort AF, Sturm MT, Schuhen K (2018a) A new approach for the agglomeration and subsequent removal of polyethylene, polypropylene, and mixtures of both from freshwater systems–a case study. Environ Sci Pollut R 25:15226–15234. https://doi.org/10.1007/s11356-018-1981-7

40. Herbort AF, Sturm MT, Fiedler S, Abkai G, Schuhen K (2018b) Alkoxy-silyl induced agglomeration: a new approach for the sustainable removal of microplastic from aquatic systems. J Polym Environ 26:4258–4270. https://doi.org/10.1007/s10924-018-1287-3
41. An C, Huang G, Yao Y, Zhao S (2017) Emerging usage of electrocoagulation technology for oil removal from wastewater: a review. Sci Total Environ 579:537–556. https://doi.org/10.1016/j.scitotenv.2016.11.062
42. Perren W, Wojtasik A, Cai Q (2018) Removal of microbeads from wastewater using electrocoagulation. ACS Omega 3:3357–3364. https://doi.org/10.1021/acsomega.7b02037
43. Skaf DW, Punzi VL, Rolle JT, Kleinberg KA (2020) Removal of micron-sized microplastic particles from simulated drinking water via alum coagulation. Chem Eng J 386:123807. https://doi.org/10.1016/j.cej.2019.123807
44. Xia Y, Xiang X, Dong K, Gong Y, Li Z (2020) Surfactant stealth effect of microplastics in traditional coagulation process observed via 3-D fluorescence imaging. Sci Total Environ 729:138783. https://doi.org/10.1016/j.scitotenv.2020.138783
45. Zhang Y, Diehl A, Lewandowski A, Gopalakrishnan K, Baker T (2020b) Removal efficiency of micro- and nanoplastics (180 nm–125 μm) during drinking water treatment. Sci Total Environ 720:137383. https://doi.org/10.1016/j.scitotenv.2020.137383
46. Shahi NK, Maeng M, Kim D, Dockko S (2020) Removal behavior of microplastics using alum coagulant and its enhancement using polyamine-coated sand. Process Saf Environ 141:9–17. https://doi.org/10.1016/j.psep.2020.05.020
47. Lapointe M, Farner JM, Hernandez LM, Tufenkji N (2020) Understanding and improving microplastic removal during water treatment: impact of coagulation and flocculation. Environ Sci Technol 54:8719–8727. https://doi.org/10.1021/acs.est.0c00712
48. Sturm MT, Herbort AF, Horn H, Schuhen K (2020) Comparative study of the influence of linear and branched alkyltrichlorosilanes on the removal efficiency of polyethylene and polypropylene-based microplastic particles from water. Environ Sci Pollut R 27:10888–10898. https://doi.org/10.1007/s11356-020-07712-9
49. Moussa DT, El-Naas MH, Nasser M, Al-Marri MJ (2017) A comprehensive review of electrocoagulation for water treatment: potentials and challenges. J Environ Manag 186:24–41. https://doi.org/10.1016/j.jenvman.2016.10.032
50. Padervand M, Lichtfouse E, Robert D, Wang C (2020) Removal of microplastics from the environment. A review. Environ Chem Lett 18:807–828. https://doi.org/10.1007/s10311-020-00983-1
51. Akarsu C, Deniz F (2021) Electrocoagulation/Electroflotation process for removal of organics and microplastics in laundry wastewater. In: Clean-Soil Air Water, vol 49, p 2000146. https://doi.org/10.1002/clen.202000146
52. Ahmed MB, Rahman MS, Alom J, Hasan MDS, Johir MAH, Ibrahim M, Mondal H, Lee DY, Park J, Zhou JL, Yoon MH (2021) Microplastic particles in the aquatic environment: a systematic review. Sci Total Environ 775:145793
53. Hidayaturrahman H, Lee TG (2019) A study on characteristics of microplastic in wastewater of South Korea: identification, quantification, and fate of microplastics during treatment process. Mar Pollut Bull 146:696–670. https://doi.org/10.1016/j.marpolbul.2019.06.071
54. Ahmed MB, Zhou JL, Ngo HH, Guo W, Thomaidis NS, Xu J (2017) Progress in the biological and chemical treatment technologies for emerging contaminant removal from wastewater: a critical review. J Hazard Mater 323:274–298. https://doi.org/10.1016/j.jhazmat.2016.04.045
55. Tagg AS, Harrison JP, Ju-Nam Y, Sapp M, Bradley EL, Sinclair CJ, Ojeda JJ (2017) Fenton's reagent for the rapid and efficient isolation of microplastics from wastewater. Chem Commun 53:372. https://doi.org/10.1039/c6cc08798a
56. Kang J, Zhou L, Duan X, Sun H, Ao Z, Wang S (2019) Degradation of cosmetic microplastics via functionalized carbon Nanosprings. Matter 1:745–758. https://doi.org/10.1016/j.matt.2019.06.004
57. Ganzenko O, Huguenot D, van Hullebusch ED, Esposito G, Oturan MA (2014) Electrochemical advanced oxidation and biological processes for wastewater treatment: a review of the

combined approaches. Environ Sci Pollut Res 21:8493–8524. https://doi.org/10.1007/s11356-014-2770-6

58. Miao F, Liu YF, Gao MM, Yu X, Xiao PW, Wang M, Wang SG, Wang XH (2020) Degradation of polyvinyl chloride microplastics via an electro-Fenton-like system with a TiO_2/graphite cathode. J Hazard Mater 399:123023. https://doi.org/10.1016/j.jhazmat.2020.123023
59. Liang W, Luo Y, Song S, Dong X, Yu X (2013) High photocatalytic degradation activity of polyethylene containing polyacrylamide grafted TiO_2. Polym Degrad Stabil 98:1754–1761. https://doi.org/10.1016/j.polymdegradstab.2013.05.027
60. Ariza-Tarazona MC, Villarreal-Chiu JF, Barbieri V, Siligardi C, Cedillo-González EI (2019) New strategy for microplastic degradation: green photocatalysis using a protein-based porous N-TiO_2 semiconductor. Ceram Int 45:9618–9624. https://doi.org/10.1016/j.ceramint.2018.10.208
61. Tofa TS, Kunjali KL, Paul S, Dutta J (2019) Visible light photocatalytic degradation of microplastic residues with zinc oxide nanorods. Environ Chem Lett 17:1341–1346. https://doi.org/10.1007/s10311-019-00859-z
62. Domínguez-Jaimes LP, Cedillo-Gonzalez EI, Luevano-Hipolito E, Acuna-Bedoya JD, Hernandez-López JM (2021) Degradation of primary nanoplastics by photocatalysis using different anodized TiO_2 structures. J Hazard Mater 413:125452. https://doi.org/10.1016/j.jhazmat.2021.125452
63. Grbic J, Nguyen B, Guo E, You JB, Sinton D, Rochman CM (2019) Magnetic extraction of microplastics from environmental samples. Environ Sci Tech Let 6:68–72. https://doi.org/10.1021/acs.estlett.8b00671
64. Rhein F, Scholl F, Nirschl H (2019) Magnetic seeded filtration for the separation of fine polymer particles from dilute suspensions: microplastics. Chem Eng Sci 207:1278–1287. https://doi.org/10.1016/j.ces.2019.07.052
65. Tang Y, Zhang S, Su Y, Wu D, Zhao Y, Xie B (2021) Removal of microplastics from aqueous solutions by magnetic carbon nanotubes. Chem Eng J 406:126804. https://doi.org/10.1016/j.cej.2020.126804
66. Mani T, Frehland S, Kalberer A, Burkhardt-Holm P (2019) Using castor oil to separate microplastics from four different environmental matrices. Anal. Methods-UK 11:1788–1794. https://doi.org/10.1039/C8AY02559B
67. Lares M, Ncibi MC, Sillanpää M, Sillanpää M (2019) Intercomparison study on commonly used methods to determine microplastics in wastewater and sludge samples. Environ Sci Pollut R 26:12109–12122. https://doi.org/10.1007/s11356-019-04584-6
68. Wang C, Chelazzi D, Mikola J, Leiniö V, Heikkinen R, Cincinelli A, Pellinen J (2020c) Olive oil-based method for the extraction, quantification and identification of microplastics in soil and compost samples. Sci Total Environ 733:139338. https://doi.org/10.1016/j.scitotenv.2020.139338
69. Wang H, Wang C, Fu J, Gu G (2013) Wetting behavior and mechanism of wetting agents on low-energy surface. Colloid Surface A 424:10–17. https://doi.org/10.1016/j.colsurfa.2013.01.063
70. Enfrin M, Dumée LF, Lee J (2019) Nano/microplastics in water and wastewater treatment processes–origin, impact and potential solutions. Water Res 161:621–638. https://doi.org/10.1016/j.watres.2019.06.049
71. Zhang M, Yang J, Kang Z, Wu X, Tang L, Qiang Z, Zhang D, Pan X (2021a) Removal of micron-scale microplastic particles from different waters with efficient tool of surface-functionalized microbubbles. J Hazard Mater 404:124095. https://doi.org/10.1016/j.jhazmat.2020.124095
72. Zhang YS, Jiang HR, Bian K, Wang H, Wang CQ (2021b) Is froth flotation a potential scheme for microplastics removal? Analysis on flotation kinetics and surface characteristics. Sci Total Environ 792:148345. https://doi.org/10.1016/j.scitotenv.2021.148345

73. Wang ZG, Sun CZ, Li FM, Chen LY (2021) Fatigue resistance, reusable and biodegradable sponge materials from plant protein with rapid water adsorption capacity for microplastics removal. Chem Eng J 415:129006. https://doi.org/10.1016/j.cej.2021.129006
74. Tiwari E, Singh N, Khandelwal N, Monikh FA, Darbha GK (2020) Application of Zn/Al layered double hydroxides for the removal of nano-scale plastic debris from aqueous systems. J Hazard Mater 397:122769. https://doi.org/10.1016/j.jhazmat.2020.122769
75. Sun C, Wang Z, Chen L, Li F (2020) Fabrication of robust and compressive chitin and graphene oxide sponges for removal of microplastics with different functional groups. Chem Eng J 393: 124796. https://doi.org/10.1016/j.cej.2020.124796
76. Chen ZY, Huang ZJ, Liu JH, Wu EY, Zheng Q, Cui LH (2021) Phase transition of mg/Al-flocs to mg/Al-layered double hydroxides during flocculation and polystyrene nanoplastics removal. J Hazard Mater 406:124697. https://doi.org/10.1016/j.jhazmat.2020.124697
77. Singh N, Khandelwal N, Ganie ZA, Tiwari E, Darbha GK (2021) Eco-friendly magnetic biochar: An effective trap for nanoplastics of varying surface functionality and size in the aqueous environment. Chem Eng J 418:129405. https://doi.org/10.1016/j.cej.2021.129405
78. Magid ASIA, Islam MS, Chen YL, Weng LP, Li JB, Ma J, Li YT (2021) Enhanced adsorption of polystyrene nanoplastics (PSNPs) onto oxidized corncob biochar with high pyrolysis temperature. Sci Total Environ 784:147115. https://doi.org/10.1016/j.scitotenv.2021.147115
79. Ganie ZA, Khandelwal N, Tiwari E, Singh N, Darbha GK (2021) Biochar-facilitated remediation of nanoplastic contaminated water: effect of pyrolysis temperature induced surface modifications. J Hazard Mater 417:126096. https://doi.org/10.1016/j.jhazmat.2021.126096
80. Chen Z, Zhao W, Xing R, Xie S, Yang X, Cui P, Lu J, Liao H, Yu Z, Wang S, Zhou S (2020) Enhanced in situ biodegradation of microplastics in sewage sludge using hyperthermophilic composting technology. J Hazard Mater 384:121271. https://doi.org/10.1016/j.jhazmat.2019.121271
81. Lange K, Magnusson K, Viklander M, Blecken G-T (2021) Removal of rubber, bitumen and other microplastic particles from stormwater by a gross pollutant trap–bioretention treatment train. Water Res 202:117457. https://doi.org/10.1016/j.watres.2021.117457
82. Summers S, Henry T, Gutierrez T (2018) Agglomeration of nano- and microplastic particles in seawater by autochthonous and de novo-produced sources of exopolymeric substances. Mar Pollut Bull 130:258–267. https://doi.org/10.1016/j.marpolbul.2018.03.039
83. Cunha C, Silva L, Paulo J, Faria M, Nogueira N, Cordeiro N (2020) Microalgal-based biopolymer for nano- and microplastic removal: a possible biosolution for wastewater treatment. Environ Pollut 263:114385. https://doi.org/10.1016/j.envpol.2020.114385

Environmental Microplastics: A Significant Pollutant of the Anthropocene

Arisha Saif Uddin, Saif Uddin, and Scott W. Fowler

1 Introduction

Plastic waste has become a menace globally. Its omnipresence and ever growing environmental concern, over the years, have generated a great deal of scientific interest. It is estimated that 8.3 billion metric tons of virgin plastics has been produced between 1950 and 2015, resulting in approximately 6.3 million metric tons of plastic waste of which around 9% is recycled, 12% incinerated, and a huge 79% ends up in landfills [1]. The earth is turning into planet plastic. The problem is further exacerbated when plastic breaks down into smaller bits, due to biotic as well as abiotic degradation, forming what is termed microplastics (MPs). Over 6500 scientific publications have been published on MPs in different environmental matrixes, most of them looking at marine waters [2], about 50 studies have observed the presence of MPs in aerosols [3] and 18 studies have reported MP in groundwater [4].

It is believed MPs are one of the largest contaminants by volume created by humans and have become an immense environmental and ecological concern in aquatic environments [5]. The origin of MP is linked to the colossal quantity of plastic entering the aquatic systems, mainly from poorly managed plastic waste disposal [6–11]. The MP in the environment occurs both as "primary" and

A. S. Uddin
Stephen Perse Foundation, Union Road, Cambridge, UK

S. Uddin (✉)
Environment and Life Sciences Research Center, Kuwait Institute for Scientific Research, Safat, Kuwait
e-mail: sdin@kisr.edu.kw

S. W. Fowler
School of Marine and Atmospheric Sciences, Stony Brook University, Stony Brook, NY, USA

Institute Bobby, Cap d'Ail, France

A. Khan et al. (eds.), *Microplastic sources, fate and solution*,
https://doi.org/10.1007/978-981-99-0695-6_5

"secondary" particles. The primary MPs are manufactured to be incorporated into other products, while those that break down from larger plastic debris to a size <5 mm are considered secondary MPs. Several studies have investigated the pathways of MP reaching the aquatic environments; a recent one [12] has highlighted the role of precipitation and aeolian transport in transporting both the primary and secondary MPs, in addition to the wastewater treatment streams [6, 7, 12–14].

The MPs occur as fibers, films, fragments, and spheres, in sizes varying from 1 to 500 μm. The lower size range of 1 μm is the smallest dimension at which a polymer can be identified using micro-Raman spectroscopy, while particles below 10–20 μm in size can be identified using micro-FTIR and micro-ATR-FTIR spectroscopy. The physical and chemical characteristics of MP can be very diverse in the environment, therefore Rochman et al. [15] suggested MP to be considered as a suite of contaminants rather than as a single contaminant.

The chronic exposure of MP in the aquatic environment poses a variety of issues in aquatic organisms such as inducing inflammation, malnutrition, changes in reproductive behavior and capacities [16–23], and toxicity caused by the polymer used for manufacturing plastic products [24–28]. The ability of MP to adsorb hydrophobic contaminants due to their extremely large surface-to-volume ratio and long residence time in the water column [29, 30] has attracted much attention [5–8, 21, 31–49]. The primary ecological concern from MP emanates from the fact that many aquatic organisms misidentify them as prey, and their continued ingestion may lead to malnutrition, as well as the transfer and subsequent incorporation of toxic chemicals [50] that might affect their physiological functioning. However, due to limited information on smaller inhalable MPs in aerosol, the data required to make a health risk assessment in humans are inadequate.

2 Microplastics in Marine Biota, Food, and Water

The contamination of oceans by MP is of concern not only because of the environmental impacts but also because they may compromise food safety and security and subsequently human health [51]. MP enters the land-based environment mostly from the wastewater treatment effluent, washing synthetic cloths (textile fibers), and personal care products (e.g., cosmetics and toothpaste) [52]. Moreover, some industrial applications, organic fertilizers, landfills leachates, roads and transport infrastructures, and some natural fibers (plant and animal origin) have also contributed to the MPs [53]. During the covid-19 pandemic, the extensive use of facemasks and personal protective equipment (PPE) kits has made a significant contribution to MP pollution [54, 55].

Microplastics are found in all environmental matrices of aquatic and terrestrial environments. A recent assessment indicates that globally treated effluent from wastewater treatment plants contributes about 1.47×10^{15} MPs per annum, while discharge of untreated wastewater annually adds another 3.85×10^{16} MPs to the aquatic environments [6]. Hence, the urgency to not only curb the influx of plastic into ecosystems but also to study further on the impacts these microplastics have on the delicate ecological balance provided by nature.

The fragmentation of large plastic debris by physical, chemical, and biological degradation results in secondary MPs. Textiles, single-use plastic bags, packaging wastes, fishing nets, and ropes are the main contributors of secondary MP [56, 57]. The main source of MPs in the aquatic environment are clearly WWTP outfalls, while different hydrodynamic processes such as coastal currents, waves, wind, and river outflow disperse these land-based MPs in the marine environment. MP is persistent in the marine environment and due to its small size, it is bioavailable to the coral, zooplankton, fish, and other marine animals and can potentially bioaccumulate in organisms at the higher trophic levels [52]. The size of MPs was more important than the shape of MPs for their ingestion in small predators by confusing them with natural prey items, especially in nonselective feeders [56]. The MP is difficult to degrade in the animal gut and 90% of orally ingested MP is excreted via animal feces [58]. However, Browne et al. [59] observed that the ingested MP could bioaccumulate and become translocated in different tissues of the mussel *Mytilus edulis*. Moreover, many studies have provided strong evidence for the effects of MP ingestion in different organisms, e.g., zooplankton [60, 61], echinoderms [62, 63], jellyfish [64], mussels [65, 66], oyster [28], crab [67], and fish [68].

MP has been shown to sorb various organic and inorganic contaminants from the surrounding water, sometimes up to 100 times more than from the sediments. This may be due to the high surface-to-volume ratio and non-polar surface that increases the sorption behavior [69]. The presence of toxic pollutants on MP is well known, hence, MPs serve as vectors for the transfer of toxic metals [70, 71], polybrominated diphenyl ethers (PDEs) [72, 73], polyaromatic hydrocarbons (PAHs) [74, 75], bisphenols, phthalates, and pesticides [76, 77]. These chemical compounds can be released in the gut due to acidic conditions while numerous models and studies have investigated this transport behavior [78]. Moreover, there has been observed a wide range of sub-lethal effects, i.e., gastrointestinal track injuries, malnutrition, growth, and reproduction issues, and those effects are more difficult to assess [79].

Among humans the uptake of MP particles can occur through ingestion and inhalation [80]. Seafood consumption represents one pathway for human MP exposure. Microplastic particles are often found concentrated in the digestive tract, and most humans consume clams, mussels, and small fish (e.g., sardines, anchovies and sprat) as whole organisms. Compa et al. [81], while studying European pilchard (*Sardina pilchardus*) and European anchovy (*Engraulis encrasicolus*) caught along the Mediterranean coast, found that 14–15% of the fish had MP or natural fibers in the gastrointestinal tract. The small fish, i.e. sprats and sardines are commonly sold as canned products, and Karami et al. [82] investigated the potential presence of micro and mesoplastic (size range 0.149–10 mm) in 20 different canned brands from 13 countries. They found that plastics were absent from the 16 canned brands, while the fish in 4 brands contained 1–3 particles per can. MPs were reported in 14 different dried marine fish species from 7 Asian countries [83]. Their study reported that MPs were observed in most dried fish and the highest count recorded was in dried fish from Japan (0.56 ± 0.03/g). The MPs <130 μm in diameter can potentially translocate into human tissue, and particle sizes <1.5 μm can penetrate capillaries [84]. However, an international scientific committee such as the Joint FAO/WHO Expert Committee on Food Additives (JECFA) has not yet evaluated the food safety concern based on MP. But some studies have assessed MP exposure, risk assessment, and the relative contribution of microplastic exposure to additives or chemicals [85].

The amount of MPs ingestion by an individual human depends on highly variable factors not only the characteristics of MPs, but also individual age, size, demographics, cultural heritage, geographical area, diet, and surrounding environment [86]. On a global average basis, it was estimated that humans ingest 0.1–5 g of MP weekly [86]. In another study, it is suggested that 1–5 anthropogenic MP may enter human consumers from the canned sardines and sprats [82]. This is far less compared to the amount of MP ingestion via mollusc consumption by the top European fish consumers (11,000 MP particles/year) [87]. In another study, the annual MP consumption range for Americans was reported as 39,000–52,000 particles depending on age and sex [88]. The human body's excretory system removes MP, likely disposing of >90% of ingested MP via feces. There have been concerns as to whether MP was possible to remain in the human body, potentially accumulating in some organs. Studies in mice have found that MP around 5 μm in diameter could stay in the intestines or reach the liver. The first evidence of MP in human placenta was reported 12 MP fragments (5–10 μm) [89], in the fetal side (5 fragments), the maternal side (4 fragments), and the chorioamniotic membranes (3 fragments). A different opinion was posed in another study which suggested that it is still not impossible that these MPs were the result of external contamination when the placentas were collected and analyzed [90]. However, there is as yet no detailed human health risk assessment study associated with MP due to a lack of toxicity and toxicokinetic data available for both MPs and nanoplastic [91].

Microplastic exposure also can present exposure to correlated chemicals. Once MPs are in the gut, they can release absorbed toxins, additives, and constituent monomers [92]. There have been several studies which have assessed most of the organic and inorganic toxic compounds associated with the MP, but few have assessed the relative contribution of MP exposure to that from additives or chemicals found in organisms, versus alternative exposure pathways [85]. The laboratory studies have shown that MP increased toxicity from the combination of MP and associated chemicals. The uptake of pollutants, additives, and MP damaged eco-physiological functions of certain organisms [93, 94]. However, it is difficult to estimate whether such toxicological effects translate to human health issues [59]. In animals, the quantity of chemicals accumulated from MP is suspected to be minimal compared to that obtained from other components of the diet. Moreover, researchers do not fully understand how MP interacts with human biological tissue even if there are some findings on individual effects, however it is very difficult to observe such effects at the population level. As one example, there is a significant association between urine Bisphenol A (BPA) levels and both cardiovascular disease and type II diabetes [95]. One of the human low-dose exposure assessments for BPA is via MP [85]; therefore, it is necessary for undertaking further research to assess the risk and effects of such MP dietary exposure. Another important consideration is that the MP in the environment is often in its degraded form, while most of the laboratory experiments consider new micro beads and rarely any other shape or environmentally recovered MPs for exposure.

Another aspect that has not attracted as much interest as the MP in the marine environment is aerosol. In a recent review a wide range of 1–1593 MPs m^{-3} of air is reported [3]. In the UK, the reported MP inventories were 2.9 and 712 $\pm$ 162 particle m^{-2} day^{-1} for Nottingham and Central London [46], whereas the concentration from an indoor setting in the Humber region, UK, was 1414 MP m^{-2} day^{-1} [96]. The MP concentration in the indoor air in Kuwait varied between 3.2 and 27.1 MPs m^{-3} [97]. We know that an average adult (with normal activity) inhales about 11–15 m^3 of air per day and if the air is contaminated with inhalable MPs then this gives us a perspective on the amount of MPs entering our pulmonary organs, and what risk they may pose can be ascertained with further detailed research in this area, which is currently non-existent. Moreover, it is well documented that 1 kg of laundry yields over 1900 MPs per wash. Indoor aerosol has much higher MP aerosol concentration as compared to outdoor air. Since more people are likely to spend approximately 70–80% of their time in confined indoor environments, MP inhalation exposure is likely to be very high.

Aerosols are being identified as a significant pathway for inhalation of MPs by humans; this information is vital for human health assessment. MPs in air are released from wear and tear, washing and drying clothing, erosion of synthetic rubber tires, deterioration of household furniture, emissions from the synthetic textile industry, emissions from vinyl chloride (VC) and polyvinyl chloride (PVC) industries, and contamination from city dust. Pretty much everywhere! In addition, daily, substantial quantities of plastics are burnt in open landfills causing volatilization of various harmful compounds that inevitably integrate into the atmospheric aerosol and are eventually inhaled by humans and animals. Considering the finite information available on this vital issue of MPs in aerosol, we are of the opinion that more effort and time need to be dedicated to understanding the impact of inhaling microplastics.

The long range transport of MP is very well documented with their presence in samples from the Pyrenees mountains in France [98], which was at least 75 miles away from the nearest city and was considered to be "pristine" and "uncontaminated" until then. The research suggested that microplastics were being transported by wind. Since MPs can travel through air and if they degrade into finer bits and fibers, these MPs are very likely to be affecting the pulmonary system in humans and animals and might be taken up as a priority research area. Quite intriguing the current sampling strategies for MP in aerosol samples do not always account for finer MPs; hence, the need to design and implement sampling strategies where both the inhalable and respirable fractions are collected for making realistic human health assessments should also be considered.

3 Importance of MPs Impacting Human Health

A plethora of literature exists on the presence of MPs in marine and aquatic biota [9, 16, 17, 23, 24, 36, 99–111]. MPs find their way into biota, usually by ingestion and inhalation (active uptake) [36, 112, 113] or adhesion, adsorption (passive uptake). Several workers have highlighted the potential hazard of plastic acting as a vector of pollutants that includes metals and, more often, hydrophobic organic chemicals (HOCs) [114]. The three major issues that emanate from this finding are how efficiently MPs transport metals and HOCs, the scale of this process on an oceanic scale, and whether MP ingestion contributes to metal, HOCs, and pharmaceuticals [115] bioaccumulation in marine organisms. These questions have attracted significant attention. The passive dosing of organic compounds from polymers to organisms has been long known to ecotoxicologists. Several laboratory investigations have shown that MPs are a vector and source of HOC and metals to marine organisms [35, 116–120].

A very pertinent debate is how significant is MP as a pollution source and vector compared to other pathways [2]. The published evidence is both in support and a negation of the hypothesis. Because secondary MP often results from physicochemical degradation of larger plastics, the question remains how much of HOCs and metals are on these degraded secondary MPs. There are no field data, hence, no clarity on these aspects. Much of the evidence on MP as a vector has come from experimental studies, where non-degraded primary MPs have been used. The other aspect that needs consideration is that most ingested MPs are defecated in hours [2], whereas a few gastropods have retained MPs in their gastrointestinal tract for 12 h and 7 days [121], and the longest reported time is 40 days [122]. The relative abundance of MP in the marine environment is very low compared to other vectors such as phytoplankton, zooplankton, and suspended particulates [2, 123–126], all of which have greater partition coefficients than most MPs [127]. Several studies have even suggested that MPs play a minor role in transport of organic contaminants [114, 117, 128]. So far there have been no field studies conducted to evaluate MPs as a transfer vector, conclusions are based on laboratory experiments that have used non-environmental concentrations. Furthermore, there are no assessments done on transfer factors and coefficients of transfer for metals or organics bound to MPs.

A significant information gap exists on likely bioaccumulation of organic contaminants and metals sorbed to MPs, and how it compares with the plankton species that can themselves biomagnify these pollutants and thus have both the absorbed and adsorbed components unlike MPs which contain only the adsorbed component. The presence of MP in table salt, drinking water, and seafood has been reported, but the data on its accumulation and contaminant transfer in both human and non-human biota is very limited. The few investigations that have been done suggest that most of the MPs ingested are excreted with the feces. Their toxicological potential is something that warrants serious consideration before MPs are taken up as priority pollutants.

The ecological concern from microplastics emanates from the fact that they can be inhaled by humans and can potentially lead to adverse health effects, such as localized inflammation [129], genotoxicity [130], and the development of oxidative stress, cytotoxicity [131], reproductive toxicity, carcinogenicity, and mutagenicity [132, 133]. Some studies suggested human health risks due to MP inhalation, most importantly, the respirable (PM_{10}) and inhalable ($PM_{2.5}$) fractions that can reach deep in the lungs and may be taken up by both macrophages and epithelial cells [134, 135]. Some studies have reported that fibrous MPs up to 250 μm in size avert the lung's clearance mechanisms [136]. Several health issues were reported, including reduced lung capacity in work-related conditions, coughing, and breathlessness [137–139]. Two recent reviews have highlighted the potential effects on human health [140, 141]. One study reported that microplastics deposition is more likely to occur in the upper airway tract (i.e., nose, mouth, throat) and can reach the gut when swallowed [142]. Whether the inhalation of MPs released from clothing, furnishings, synthetic tires, and degraded plastics, among other causes [37, 98, 142–150] can be a significant pathway for transferring MPs to humans is a topic which has been less explored, making it prudent to look at the finer inhalable MPs [151]. The plastic additives, dyes, and pigments could lead to over 4000 chemicals that are currently used in the plastic food packaging industry itself [152] and can provide a scale of the chemical toxicity they can induce. Most of the additives added to the plastic during processing are of small molecular size and often not chemically bound to the polymeric materials, which make them susceptible to leaching into the surrounding environment [131]. Considering the lower size cutoff for microplastics, i.e. 1 μm [153], the fine MPs have greater potential to be transferred via aerosolization into the human respiratory system [98].

4 What Do We Know About Smaller MPs and Nanoplastics

There is a significant chemical and physical diversity among the MPs present in the environment. In spite of the extremely large number of studies conducted on MPs, information on smaller MPs is quite limited, mainly due to technological challenges in sampling, processing, and identification. Another new term that has come up in a few publications is nanoplastic (NP), with a widely acceptable size range between 1 nm and 100 nm [154]. Still the amount of NPs in the environment is not known, mainly due to technological limitations; however, significant efforts have been made by undertaking laboratory experiments to assess the toxicity of NP to biota [155–185]. Most of these experiments are conducted at much higher concentrations (non-environmental). Sometimes the information is misrepresented as well, a recent review in toxicological research of nanoplastics in environment [186] reported "nanoplastics can act as carriers of chemical and biological contaminants"; however, all of the three references used to support the statement dealt with microplastics [114, 187, 188]. This is not the only case, there are hundreds of such misreportings in MP-NP literature where authors have conveniently and interchangeably used debris,

MP and NP, although they are very different things and require an entirely different approach to assessment. There are enormous technological challenges associated with NP sampling, sample preparation, identification, and characterization. Although their presence in the environment is quite significant due to their use in cosmetics and personal care products, some workers have suggested that decomposition of MP will continue to form NP [189] as well.

5 Conclusion

Microplastics is ubiquitously found in all environmental matrixes including seafood and aerosol. A large number of studies have covered the marine environment but very little effort has been directed toward examining MPs in aerosol. The risk assessments from MPs are based on experiments that essentially have not been conducted using environmental concentrations and also usually use non-degraded MPs that rarely exist in the environment. Another very pertinent point is that the risk assessments should consider both the ingested and inhaled MPs. The data on MPs in aerosol are very limited and there are practically none for size-fractionated aerosol. The smaller MPs of <1.5 μm can cross the cell barrier, there are hardly any sampling strategies to capture these ultrafine MPs, certainly drawing attention to the fact that harmonized methodologies need to be developed for capturing these smaller MPs.

Another conspicuous observation is the significant attention given to MPs as the major contaminant of the Anthropocene, it certainly is in terms of quantity but their toxicity to human and non-human biota needs extensive assessments before the present focus of science and research funding is further diverted toward MPs. There is a need to assess efficacy of MPs as vector of pollutant (organic and metals) transfer and compare it with the lower trophic level biota that are far more important transfer vectors in food chains than MPs in aquatic environment. Similar is situation with microbes in aerosol. The phytoplankton, zooplankton, and microbes are capable of biomagnifying these organic contaminants and metals, probably being more efficient vectors of pollutant transport than MPs.

Microplastic levels and their associated co-contaminants in seafood species and their resultant transfer to and consumption by humans also are not well studied in the perspective and context of global seafood security and sustainability. The claims of translocation of ingested and inhaled MPs across organs and tissues need to be underpinned with more scientific evidence. It will be interesting to explore the efficacy of MPs as vector of pollutant transfer. It is prudent to assess differences in adsorption potential of different degraded MPs. Since most of the current experimental data comes from laboratory experiments using fresh non-degraded MPs, future studies should consider these suggestions to provide a strong basis to assess the overall impact of MP on food security.

References

1. Geyer R, Jambeck JR, Law KL (2017) Production, use, and fate of all plastics ever made. Sci Adv 3(7):e1700782
2. Uddin S et al (2022a) Micro-Nano plastic in the aquatic environment: methodological problems and challenges. Animals 12(3):297
3. Habibi N et al (2022) Microplastics in the atmosphere: a review. J Environ Exposure Assess 1(1):6
4. Khant NA, Kim H (2022) Review of current issues and management strategies of microplastics in groundwater environments. Water 14(7):1020
5. Alimba CG, Faggio C (2019) Microplastics in the marine environment: current trends in environmental pollution and mechanisms of toxicological profile. Environ Toxicol Pharmacol 68:61–74
6. Uddin S, Fowler SW, Behbehani M (2020a) An assessment of microplastic inputs into the aquatic environment from wastewater streams. Mar Pollut Bull 160:111538
7. Naji A et al (2021) Microplastics in wastewater outlets of Bandar Abbas city (Iran): a potential point source of microplastics into the Persian Gulf. Chemosphere 262:128039
8. Uddin S et al (2021) A review of microplastic distribution in sediment profiles. Mar Pollut Bull 163:111973
9. Alimi OS et al (2018) Microplastics and nanoplastics in aquatic environments: aggregation, deposition, and enhanced contaminant transport. Environ Sci Technol 52:1704
10. Eerkes-Medrano D, Thompson RC, Aldridge DC (2015) Microplastics in freshwater systems: a review of the emerging threats, identification of knowledge gaps and prioritisation of research needs. Water Res 75:63
11. Habib R, Thiemann T, Kendi R (2020) Microplastics and wastewater treatment plants—a review. J Water Resour Protect 12:1–35
12. Freeman S et al (2020) Between source and sea: the role of wastewater treatment in reducing marine microplastics. J Environ Manag 266:110642
13. Sun J et al (2019) Microplastics in wastewater treatment plants: detection, occurrence and removal. Water Res 152:21–37
14. Sundt P, Schulze P-E, Syversen F (2014) Sources of microplastics-pollution to the marine environment, vol 86. Report no. M-321|2015 Norwegian environment agency, Oslo, p 20
15. Rochman CM et al (2019) Rethinking microplastics as a diverse contaminant suite. Environ Toxicol Chem 38(4):703–711
16. Abbasi S et al (2018) Microplastics in different tissues of fish and prawn from the Musa estuary Persian Gulf. Chemosphere 205:80–87
17. Barboza LGA et al (2020) Microplastic in wild fish from north East Atlantic Ocean and its potential for causing neurotoxic effects, lipid oxidative damage, and human health risks associated with ingestion exposure. Sci Total Environ 717:134625
18. Besseling E et al (2017) Fate of nano- and microplastic in freshwater systems: a modeling study. Environ Pollut 220:540–548
19. Browne MA et al (2008) Ingested microscopic plastic translocates to the circulatory system of the mussel, Mytilus edulis (L). Environ Sci Technol 42(13):5026
20. Mohsen M et al (2019) Microplastic ingestion by the farmed sea cucumber Apostichopus japonicus in China. Environ Pollut 245:1071–1078
21. Vroom RJE et al (2017) Aging of microplastics promotes their ingestion by marine zooplankton. Environ Pollut 231(Pt 1):987–996
22. Zhang S et al (2019) Microplastics in the environment: a review of analytical methods, distribution, and biological effects. TrAC Trends Anal Chem 111:62–72
23. Zakeri M et al (2020) Microplastic ingestion in important commercial fish in the southern Caspian Sea. Mar Pollut Bull 160:111598

24. Devriese LI et al (2015) Microplastic contamination in brown shrimp (Crangon crangon, Linnaeus 1758) from coastal waters of the southern North Sea and channel area. Mar Pollut Bull 98(1–2):179–187
25. Li JN et al (2019) Using mussel as a global bioindicator of coastal microplastic pollution. Environ Pollut 244:522
26. Lusher AL et al (2017) Sampling, isolating and identifying microplastics ingested by fish and invertebrates. Anal Methods 9(9):1346
27. Rochman CM et al (2013) Ingested plastic transfers hazardous chemicals to fish and induces hepatic stress. Sci Rep 3:3263
28. Sussarellu R et al (2016) Oyster reproduction is affected by exposure to polystyrene microplastics. Proc Natl Acad Sci 113(9):2430–2435
29. Ferreira I et al (2019) Nanoplastics and marine organisms: what has been studied? Environ Toxicol Pharmacol 67:1–7
30. Pan Z et al (2019) Environmental implications of microplastic pollution in the Northwestern Pacific Ocean. Mar Pollut Bull 146:215–224
31. Andrady AL (2011) Microplastics in the marine environment. Mar Pollut Bull 62(8): 1596–1605
32. Andrady AL (2017) The plastic in microplastics: a review. Mar Pollut Bull 119(1):12–22
33. Auta HS, Emenike CU, Fauziah SH (2017) Distribution and importance of microplastics in the marine environment: a review of the sources, fate, effects, and potential solutions. Environ Int 102:165–176
34. Barboza LGA et al (2019) Microplastics pollution in the marine environment, in world seas: An environmental evaluation. In: Sheppard C (ed) Volume III: ecological issues and environmental impacts. Academic Press (Elsevier), London, pp 329–351
35. Brennecke D et al (2016) Microplastics as vector for heavy metal contamination from the marine environment. Estuar Coast Shelf Sci 178:189–195
36. de Sa LC et al (2018) Studies of the effects of microplastics on aquatic organisms: what do we know and where should we focus our efforts in the future? Sci Total Environ 645:1029–1039
37. Dris R et al (2016) Synthetic fibers in atmospheric fallout: a source of microplastics in the environment? Mar Pollut Bull 104(1–2):290–293
38. Enyoh CE et al (2019) Airborne microplastics: a review study on method for analysis, occurrence, movement and risks. Environ Monit Assess 191(11):668
39. Fahrenfeld NL et al (2019) Source tracking microplastics in the freshwater environment. TrAC Trends Anal Chem 112:248–254
40. Hidalgo-Ruz V et al (2012) Microplastics in the marine environment: a review of the methods used for identification and quantification. Environ Sci Technol 46(6):3060–3075
41. Lusher A (2015) Microplastics in the marine environment: distribution, interactions and effects. In: Bergmann M, Gutow L, Klages M (eds) Marine anthropogenic litter. Springer International Publishing, Cham, pp 245–307
42. Maes T et al (2019) Shades of grey: marine litter research developments in Europe. Mar Pollut Bull 146:274–281
43. Rodrigues JP et al (2019) Significance of interactions between microplastics and POPs in the marine environment: a critical overview. TrAC Trends Anal Chem 111:252–260
44. Schymanski D et al (2018) Analysis of microplastics in water by micro-Raman spectroscopy: release of plastic particles from different packaging into mineral water. Water Res 129:154–162
45. Allen S et al (2019) Atmospheric transport and deposition of microplastics in a remote mountain catchment. Nat Geosci 12:339–344
46. Zhang Y et al (2020) Atmospheric microplastics: a review on current status and perspectives. Earth Sci Rev 203:103118
47. Wagner S et al (2018) Tire wear particles in the aquatic environment–a review on generation, analysis, occurrence, fate and effects. Water Res 139:83

48. Uddin S, Fowler SW, Saeed T (2020b) Microplastic particles in the Persian/Arabian gulf–a review on sampling and identification. Mar Pollut Bull 154:111100
49. Uddin S et al (2020c) Standardized protocols for microplastics determinations in environmental samples from the Gulf and marginal seas. Mar Pollut Bull 158:111374
50. Prinz N, Korez Š (2020) Understanding how microplastics affect marine biota on the cellular level is important for assessing ecosystem function: a review. In: Jungblut S, Liebich V, Bode-Dalby M (eds) YOUMARES 9–the oceans: our research, our future: proceedings of the 2018 conference for YOUng MArine RESearcher in Oldenburg, Germany. Springer International Publishing, Cham, pp 101–120
51. Barboza LGA et al (2018) Marine microplastic debris: An emerging issue for food security, food safety and human health. Mar Pollut Bull 133:336–348
52. Sharma S, Chatterjee S (2017) Microplastic pollution, a threat to marine ecosystem and human health: a short review. Environ Sci Pollut Res 24(27):21530–21547
53. da Costa JP et al (2018) Microplastics in soils: assessment, analytics and risks. Environ Chem 16(1):18–30
54. Ray SS et al (2022) Microplastics waste in environment: a perspective on recycling issues from PPE kits and face masks during the COVID-19 pandemic. Environ Technol Innov 26: 102290
55. Shen M et al (2021) Neglected microplastics pollution in global COVID-19: disposable surgical masks. Sci Total Environ 790:148130
56. Lehtiniemi M et al (2018) Size matters more than shape: ingestion of primary and secondary microplastics by small predators. Food Webs 17:e00097
57. Coyle R, Hardiman G, Driscoll KO (2020) Microplastics in the marine environment: a review of their sources, distribution processes, uptake and exchange in ecosystems. Case Stud Chem Environ Eng 2:100010
58. Alprol AE, Gaballah MS, Hassaan MA (2021) Micro and Nanoplastics analysis: focus on their classification, sources, and impacts in marine environment. Reg Stud Mar Sci 42:101625
59. Browne MA (2008) Environmental consequences of microplastic in marine habitats
60. Van Colen C et al (2020) Does microplastic ingestion by zooplankton affect predator-prey interactions? An experimental study on larviphagy. Environ Pollut 256:113479
61. Cole M et al (2015) The impact of polystyrene microplastics on feeding, function and fecundity in the marine copepod Calanus helgolandicus. Environ Sci Technol 49(2): 1130–1137
62. Richardson CR et al (2021) Microplastic ingestion induces asymmetry and oxidative stress in larvae of the sea urchin Pseudechinus huttoni. Mar Pollut Bull 168:112369
63. Kaposi KL et al (2014) Ingestion of microplastic has limited impact on a marine larva. Environ Sci Technol 48(3):1638–1645
64. Rapp J et al (2021) Microplastic ingestion in jellyfish Pelagia noctiluca (Forsskal, 1775) in the North Atlantic Ocean. Mar Pollut Bull 166:112266
65. Patterson J et al (2021) Microplastic contamination in Indian edible mussels (Perna perna and Perna viridis) and their environs. Mar Pollut Bull 171:112678
66. von Moos N, Burkhardt-Holm P, Köhler A (2012) Uptake and effects of microplastics on cells and tissue of the blue mussel Mytilus edulis L. after an experimental exposure. Environ Sci Technol 46(20):11327–11335
67. Watts AJR et al (2014) Uptake and retention of microplastics by the shore crab Carcinus maenas. Environ Sci Technol 48(15):8823–8830
68. Batel A et al (2016) Transfer of benzo[a]pyrene from microplastics to Artemia nauplii and further to zebrafish via a trophic food web experiment: CYP1A induction and visual tracking of persistent organic pollutants. Environ Toxicol Chem 35(7):1656–1666
69. Rochman CM, Hentschel BT, Teh SJ (2014) Long-term sorption of metals is similar among plastic types: implications for plastic debris in aquatic environments. PLoS One 9(1):e85433
70. Liu S et al (2022) Microplastics as a vehicle of heavy metals in aquatic environments: a review of adsorption factors, mechanisms, and biological effects. J Environ Manag 302:113995

71. Hildebrandt L et al (2021) Microplastics as a Trojan horse for trace metals. J Hazard Mater Lett 2:100035
72. Sun B et al (2021) Leaching of polybrominated diphenyl ethers from microplastics in fish oil: kinetics and bioaccumulation. J Hazard Mater 406:124726
73. Singla M et al (2020) Sorption and release process of polybrominated diphenyl ethers (PBDEs) from different composition microplastics in aqueous medium: solubility parameter approach. Environ Pollut 262:114377
74. Peng B et al (2022) Assessment and sources identification of microplastics, PAHs and OCPs in the Luoyuan Bay, China: based on multi-statistical analysis. Mar Pollut Bull 175:113351
75. Akhbarizadeh R et al (2021) Suspended fine particulate matter (PM2.5), microplastics (MPs), and polycyclic aromatic hydrocarbons (PAHs) in air: their possible relationships and health implications. Environ Res 192:110339
76. Montoto-Martínez T et al (2021) Microplastics, bisphenols, phthalates and pesticides in odontocete species in the Macaronesian region (eastern North Atlantic). Mar Pollut Bull 173:113105
77. Li H et al (2021) Adsorption of three pesticides on polyethylene microplastics in aqueous solutions: kinetics, isotherms, thermodynamics, and molecular dynamics simulation. Chemosphere 264:128556
78. Carbery M, O'Connor W, Palanisami T (2018a) Trophic transfer of microplastics and mixed contaminants in the marine food web and implications for human health. Environ Int 115:400–409
79. Panti C et al (2019) Marine litter: one of the major threats for marine mammals. Outcomes from the European cetacean society workshop. Environ Pollut 247:72–79
80. Danopoulos E et al (2022) A rapid review and meta-regression analyses of the toxicological impacts of microplastic exposure in human cells. J Hazard Mater 427:127861
81. Compa M et al (2018) Ingestion of microplastics and natural fibres in Sardina pilchardus (Walbaum, 1792) and Engraulis encrasicolus (Linnaeus, 1758) along the Spanish Mediterranean coast. Mar Pollut Bull 128:89–96
82. Karami A et al (2018) Microplastic and mesoplastic contamination in canned sardines and sprats. Sci Total Environ 612:1380–1386
83. Piyawardhana N et al (2022) Occurrence of microplastics in commercial marine dried fish in Asian countries. J Hazard Mater 423:127093
84. Lundebye A-K, Lusher AL, M.S. Bank (2022) Marine microplastics and seafood: implications for food security. In: M.S. Bank (ed) Microplastic in the environment: pattern and process. Springer International, Cham, pp 131–153
85. Smith M et al (2018) Microplastics in seafood and the implications for human health. Curr Environ Health Rep 5(3):375–386
86. Senathirajah K et al (2021) Estimation of the mass of microplastics ingested–a pivotal first step towards human health risk assessment. J Hazard Mater 404:124004
87. Van Cauwenberghe L, Janssen CR (2014) Microplastics in bivalves cultured for human consumption. Environ Pollut 193:65–70
88. Cox KD et al (2019a) Human consumption of microplastics. Environ Sci Technol 53:7068
89. Ragusa A et al (2021) Plasticenta: first evidence of microplastics in human placenta. Environ Int 146:106274
90. Lim X (2021) Microplastics Are Everywhere—But Are They Harmful? Nature 593:22–25
91. Panel E, Chain F (2016) Presence of microplastics and nanoplastics in food, with particular focus on seafood. EFSA J 14(6):e04501
92. Cox KD et al (2019b) Human consumption of microplastics. Environ Sci Technol 53(12): 7068–7074
93. Browne MA et al (2013) Microplastic moves pollutants and additives to Worms, reducing functions linked to health and biodiversity. Curr Biol 23(23):2388–2392
94. Gallo F et al (2018) Marine litter plastics and microplastics and their toxic chemicals components: the need for urgent preventive measures. Environ Sci Eur 30(1):13–13

95. Provvisiero DP et al (2016) Influence of bisphenol a on type 2 diabetes mellitus. Int J Environ Res Public Health 13(10):989
96. Jenner LC et al (2021) Household indoor microplastics within the Humber region (United Kingdom): quantification and chemical characterisation of particles present. Atmos Environ 259:118512
97. Uddin S et al (2022b) A preliminary assessment of size-fractionated microplastics in indoor aerosol-Kuwait's baseline. Toxics 10(2):71
98. Uddin S, Habibi N, Fowler SW, Behbehani M, Gevao B, Faizuddin M, Gorgun AU (2023) Aerosols as vectors for contaminants: a perspective based on outdoor aerosol data from Kuwait. Atmosphere 14:470. https://doi.org/10.3390/atmos14030470
99. Akhbarizadeh R, Moore F, Keshavarzi B (2018) Investigating a probable relationship between microplastics and potentially toxic elements in fish muscles from northeast of Persian gulf. Environ Pollut 232:154–163
100. Akhbarizadeh R, Moore F, Keshavarzi B (2019) Investigating microplastics bioaccumulation and biomagnification in seafood from the Persian Gulf: a threat to human health? Food Addit Contam Part A Chem Anal Control Expo Risk Assess 36(11):1696–1708
101. Al-Lihaibi S et al (2019) Microplastics in sediments and fish from the Red Sea coast at Jeddah (Saudi Arabia). Environ Chem 16(8):641–650
102. Alomar C et al (2017) Microplastic ingestion by Mullus surmuletus (Linnaeus, 1758) fish and its potential for causing oxidative stress. Environ Res 159:135–142
103. Al-Salem SM, Uddin S, Lyons B (2020) Evidence of microplastics (MP) in gut content of major consumed marine fish species in the State of Kuwait (of the Arabian/Persian gulf). Mar Pollut Bull 154:111052
104. Alvarez G, Barros A, Velando A (2018) The use of European shag pellets as indicators of microplastic fibers in the marine environment. Mar Pollut Bull 137:444–448
105. Bagheri T et al (2020) Microplastics distribution, abundance and composition in sediment, fishes and benthic organisms of the Gorgan Bay Caspian Sea. Chemosphere 257:127201
106. Besseling E et al (2013) Effects of microplastic on fitness and PCB bioaccumulation by the lugworm Arenicola marina (L.). Environ Sci Technol 47(1):593
107. Carbery M, O'Connor W, Palanisami T (2018b) Trophic transfer of microplastics and mixed contaminants in the marine food web and implications for human health. Environ Int 115:400
108. Catarino AI et al (2018) Low levels of microplastics (MP) in wild mussels indicate that MP ingestion by humans is minimal compared to exposure via household fibres fallout during a meal. Environ Pollut 237:675–684
109. Collard F et al (2017) Microplastics in livers of European anchovies (Engraulis encrasicolus, L.). Environ Pollut 229:1000–1005
110. de Sa LC, Luís LG, Guilhermino L (2015) Effects of microplastics on juveniles of the common goby (Pomatoschistus microps): confusion with prey, reduction of the predatory performance and efficiency, and possible influence of developmental conditions. Environ Pollut 196:359–361
111. Maass S et al (2017) Transport of microplastics by two collembolan species. Environ Pollut 225:456
112. Hermsen W, Sims I, Crane M (1994) The bioavailability and toxicity to Mytilus edulis L. of two organochlorine pesticides adsorbed to suspended solids. Mar Environ Res 38:61–69
113. Kühn S et al (2015) Marine anthropogenic litter, p 75
114. Koelmans AA et al (2016) Microplastic as a vector for Chemicals in the Aquatic Environment: critical review and model-supported reinterpretation of empirical studies. Environ Sci Technol 50(7):3315–3326
115. Yao J et al (2022) Surface functional groups determine adsorption of pharmaceuticals and personal care products on polypropylene microplastics. J Hazard Mater 423(Pt B):127131
116. Jinadasa BKKK, Uddin S, Fowler SW (2023) Microplastics (MPs) in marine food chains: is it a food safety issue? Adv Food Nutr Res 103:101–140. https://doi.org/10.1016/bs.afnr.2022.07.005

117. Gouin T et al (2011) Thermodynamic approach for assessing the environmental exposure of chemicals absorbed to microplastic. Environ Sci Technol 45:1466–1472
118. Chua EM et al (2014) Assimilation of Polybrominated diphenyl ethers from microplastics by the marine amphipod, Allorchestes Compressa. Environ Sci Technol 48:8127–8134
119. Abbasi S et al (2020) PET-microplastics as a vector for heavy metals in a simulated plant rhizosphere zone. Sci Total Environ 744:140984
120. Gao F et al (2019) Study on the capability and characteristics of heavy metals enriched on microplastics in marine environment. Mar Pollut Bull 144:61–67
121. Weber A et al (2021) Ingestion and toxicity of microplastics in the freshwater gastropod Lymnaea stagnalis: no microplastic-induced effects alone or in combination with copper. Chemosphere 263:128040
122. Duis K, Coors A (2016) Microplastics in the aquatic and terrestrial environment: sources (with a specific focus on personal care products), fate and effects. Environ Sci Eur 28(1):2
123. Uddin S et al (2018a) ^{210}Po concentration in selected diatoms and dinoflagellates in the northern Arabian gulf. Mar Pollut Bull 129(1):343–346
124. Uddin S et al (2018b) ^{210}Po concentration in selected calanoid copepods in the northern Arabian gulf. Mar Pollut Bull 133:861–864
125. Uddin S et al (2019) Concentration of 210Po and 210Pb in macroalgae from the northern gulf. Mar Pollut Bull 145:474–479
126. Annabi-Trabelsi N et al (2021) Concentrations of trace metals in phytoplankton and zooplankton in the Gulf of Gabès, Tunisia. Mar Pollut Bull 168:112392
127. Mato Y et al (2001) Plastic resin pellets as a transport medium for toxic chemicals in the marine environment. Environ Sci Technol 35(2):318
128. Koelmans AA et al (2013) Correction to plastic as a carrier of POPs to aquatic organisms. A model analysis. Environ Sci Technol 47:8992–8993
129. Wright SL, Kelly FJ (2017) Plastic and human health: a micro issue? Environ Sci Technol 51(12):6634–6647
130. Gasperi J et al (2018) Microplastics in air: are we breathing it in? Curr Opin Environ Sci Health 1:1–5
131. Prata J (2018) Airborne microplastics: consequences to human health? Environ Pollut (Barking, Essex: 1987) 234:115–126
132. Marzec A (2014) The effect of dyes, pigments and ionic liquids on the properties of elastomer composites. Université Claude Bernard - Lyon I and Uniwersytet lódzki, Lodz. 175p
133. Campanale C et al (2020) A detailed review study on potential effects of microplastics and additives of concern on human health. Int J Environ Res Public Health 17(4):1212
134. Schulze F et al (2017) Air quality effects on human health and approaches for its assessment through microfluidic chips. Genes 8(10):244
135. Xing Y-F et al (2016) The impact of PM2.5 on the human respiratory system. J Thorac Dis 8(1):E69–E74
136. Pauly JL et al (1998) Inhaled cellulosic and plastic fibers found in human lung tissue. Cancer Epidemiol Biomark Prev 7:419
137. Boag AH et al (1999) The pathology of interstitial lung disease in nylon flock workers. Am J Surg Pathol 23(12):1539
138. Eschenbacher WL et al (1999) Nylon flock associated interstitial lung disease. Am J Respir Crit Care Med 159(6):2003
139. Kremer AM et al (1994) Airway hyper-responsiveness and the prevalence of work-related symptoms in workers exposed to irritants. Am J Ind Med 26(5):655
140. Amato-Lourenço LF et al (2020) An emerging class of air pollutants: potential effects of microplastics to respiratory human health? Sci Total Environ 749:141676–141676
141. Sridharan S et al (2021) Microplastics as an emerging source of particulate air pollution: a critical review. J Hazard Mater 418:126245
142. Wright S, Levermore J, Kelly F (2019) Raman spectral imaging for the detection of inhalable microplastics in ambient particulate matter samples. Environ Sci Technol 53(15):8947–8956

143. Uddin S, Fowler SW, Behbehani M (2023) 210Po in the environment: reassessment of dose to humans. Sustainability 15(2):1674
144. Abbasi S et al (2019) Distribution and potential health impacts of microplastics and microrubbers in air and street dusts from Asaluyeh County. Iran Environ Pollut 244:153–164
145. Dris R et al (2015) Microplastic contamination in an urban area: a case study in greater Paris. Environ Chem 12(5):592–599
146. Dris R et al (2017) A first overview of textile fibers, including microplastics, in indoor and outdoor environments. Environ Pollut 221:453–458
147. Dris R, Gasperi J, Tassin B (2018) Sources and fate of microplastics in urban areas: a focus on Paris megacity. In: Wagner M, Lambert S (eds) Freshwater microplastics: emerging environmental contaminants? Springer, Cham, pp 69–83
148. Wright SL et al (2020) Atmospheric microplastic deposition in an urban environment and an evaluation of transport. Environ Int 136:105411
149. Napper IE, Thompson RC (2016) Release of synthetic microplastic plastic fibres from domestic washing machines: effects of fabric type and washing conditions. Mar Pollut Bull 112(1):39–45
150. Wright SL et al (2021) Development of screening criteria for microplastic particles in air and atmospheric deposition: critical review and applicability towards assessing human exposure. Microplastics Nanoplastics 1(1):6
151. Bianco A, Passananti M (2020) Atmospheric micro and Nanoplastics: An enormous microscopic problem. Sustainability 12(18):7327
152. Zimmermann L et al (2019) Benchmarking the in vitro toxicity and chemical composition of plastic consumer products. Environ Sci Technol 53(19):11467–11477
153. Hartmann NB et al (2019) Are we speaking the same language? Recommendations for a definition and categorization framework for plastic debris. Environ Sci Technol 53(3): 1039–1047
154. Jambeck JR et al (2015) Plastic waste inputs from land into the ocean. Science 347(6223): 768–771
155. Bhattacharjee S et al (2014) Role of membrane disturbance and oxidative stress in the mode of action underlying the toxicity of differently charged polystyrene nanoparticles. RSC Adv 4(37):19321–19330
156. Torres-Ruiz M et al (2021) Toxicity of nanoplastics for zebrafish embryos, what we know and where to go next. Sci Total Environ 797:149125
157. Amereh F et al (2020) The emerging risk of exposure to nano(micro)plastics on endocrine disturbance and reproductive toxicity: from a hypothetical scenario to a global public health challenge. Environ Pollut 261:114158
158. Amereh F et al (2019) Thyroid endocrine status and biochemical stress responses in adult male Wistar rats chronically exposed to pristine polystyrene nanoplastics. Toxicol Res 8(6): 953–963
159. Besseling E et al (2019) Quantifying ecological risks of aquatic micro- and nanoplastic. Crit Rev Environ Sci Technol 49(1):32–80
160. Bhagat J, Nishimura N, Shimada Y (2021) Toxicological interactions of microplastics/ nanoplastics and environmental contaminants: current knowledge and future perspectives. J Hazard Mater 405:123913
161. Bhagat J et al (2020) Zebrafish: An emerging model to study microplastic and nanoplastic toxicity. Sci Total Environ 728:138707
162. Brandts I et al (2020) Polystyrene nanoplastics accumulate in ZFL cell lysosomes and in zebrafish larvae after acute exposure, inducing a synergistic immune response: in vitro without affecting larval survival in vivo. Environ Sci Nano 7(8):2410–2422
163. Brun NR et al (2018) Nanoparticles induce dermal and intestinal innate immune system responses in zebrafish embryos. Environ Sci Nano 5(4):904–916
164. Brun NR et al (2019) Polystyrene nanoplastics disrupt glucose metabolism and cortisol levels with a possible link to behavioural changes in larval zebrafish. Commun Biol **2**(1):382

165. Catarino AI, Frutos A, Henry TB (2019) Use of fluorescent-labelled nanoplastics (NPs) to demonstrate NP absorption is inconclusive without adequate controls. Sci Total Environ 670: 915–920
166. Chae Y, An YJ (2017) Effects of micro- and nanoplastics on aquatic ecosystems: current research trends and perspectives. Mar Pollut Bull 124(2):624–632
167. Chang X et al (2020) Potential health impact of environmental micro- and nanoplastics pollution. J Appl Toxicol 40(1):4–15
168. Chen Q et al (2017) Quantitative investigation of the mechanisms of microplastics and nanoplastics toward zebrafish larvae locomotor activity. Sci Total Environ 584-585:1022–1031
169. Davranche M et al (2020) Nanoplastics on the coast exposed to the North Atlantic gyre: evidence and traceability. NanoImpact 20:100262
170. Duan Z et al (2020) Barrier function of zebrafish embryonic chorions against microplastics and nanoplastics and its impact on embryo development. J Hazard Mater 395:122621
171. Fadare OO et al (2020) Eco-Corona vs protein Corona: effects of humic substances on Corona formation and Nanoplastic particle toxicity in daphnia magna. Environ Sci Technol 54(13): 8001–8009
172. Gaylarde CC, Baptista Neto JA, da Fonseca EM (2021) Nanoplastics in aquatic systems - are they more hazardous than microplastics? Environ Pollut 272:115950
173. Heinlaan M et al (2020) Hazard evaluation of polystyrene nanoplastic with nine bioassays did not show particle-specific acute toxicity. Sci Total Environ 707:136073
174. Hu Q et al (2021) Polystyrene nanoparticles trigger the activation of p38 MAPK and apoptosis via inducing oxidative stress in zebrafish and macrophage cells. Environ Pollut 269:116075
175. Jeon S et al (2018) Surface charge-dependent cellular uptake of polystyrene nanoparticles. Nano 8(12):1028
176. Ji Y et al (2020) Realistic polyethylene terephthalate nanoplastics and the size- and surface coating-dependent toxicological impacts on zebrafish embryos. Environ Sci Nano 7(8): 2313–2324
177. Kik K, Bukowska B, Sicińska P (2020) Polystyrene nanoparticles: sources, occurrence in the environment, distribution in tissues, accumulation and toxicity to various organisms. Environ Pollut 262:114297
178. Kögel T et al (2020) Micro- and nanoplastic toxicity on aquatic life: determining factors. Sci Total Environ 709:136050
179. Lee WS et al (2019) Bioaccumulation of polystyrene nanoplastics and their effect on the toxicity of au ions in zebrafish embryos. Nanoscale 11(7):3200–3207
180. Lehner R et al (2019) Emergence of Nanoplastic in the environment and possible impact on human health. Environ Sci Technol 53(4):1748–1765
181. Liu Y et al (2021a) Effects of Nanoplastics and butyl Methoxydibenzoylmethane on early zebrafish embryos identified by single-cell RNA sequencing. Environ Sci Technol 55(3): 1885–1896
182. Liu Y et al (2019) Ecotoxicological effects on Scenedesmus obliquus and Danio rerio co-exposed to polystyrene nano-plastic particles and natural acidic organic polymer. Environ Toxicol Pharmacol 67:21–28
183. Liu Z et al (2020) Effects of nanoplastics at predicted environmental concentration on Daphnia pulex after exposure through multiple generations. Environ Pollut 256:113506
184. Liu Z et al (2021b) Development of an adverse outcome pathway for nanoplastic toxicity in Daphnia pulex using proteomics. Sci Total Environ 766:144249
185. Mattsson K et al (2017) Brain damage and behavioural disorders in fish induced by plastic nanoparticles delivered through the food chain. Sci Rep 7(1):11452

186. Shen M et al (2019) Recent advances in toxicological research of nanoplastics in the environment: a review. Environ Pollut 252:511–521
187. Hodson ME et al (2017) Plastic bag derived-microplastics as a vector for metal exposure in terrestrial invertebrates. Environ Sci Technol 51(8):4714–4721
188. Laganà P et al (2019) Do plastics serve as a possible vector for the spread of antibiotic resistance? First insights from bacteria associated to a polystyrene piece from King George Island (Antarctica). Int J Hyg Environ Health 222:89–100
189. Mattsson K et al (2018) Chapter 13–Nanoplastics in the aquatic environment. In: Zeng EY (ed) Microplastic contamination in aquatic environments. Elsevier, pp 379–399

Conveyance, Bounty, and Dangers of Microplastics in Nature

Fabiana Corami and Beatrice Rosso

1 Introduction

Plastic debris is ubiquitous in the environment resulting in potential threats to ecosystems, biota, or even human health. With global production and rapid expansion, the amount of anthropogenic plastic litter in all environmental matrices has increased dramatically over the last few years. Plastics started to reach the market at a large scale in the 1950s; then, the global production of plastic dramatically increased from 0.5 million tons (Mt) per year in 1960 to 359 million tons [1]. Because of its versatile properties, the demand for plastics rapidly grows for various applications: transport, textiles, building, construction, medical and packaging materials. Hence, thousands of plastic goods are demanded in the customer market in our everyday lives, and over 6300 Mt of plastic waste has been produced worldwide. Only 9% of plastic is recycled, 12% is burned, and 79% is accumulated in the environment [2].

Plastic pollution in all environmental matrices has been extensively reported [3, 4] and thus attracted increasing public, scientific, and political attention. Only recently, environmental occurrences of microplastics (MPs), smaller pieces of plastic, have been detected frequently, and they interact with all the environmental systems.

The term "microplastics," coined in 2004 [5], was applied to define the smaller plastic particles in marine sediments and the water column of European waters for the first time. An explicit agreement on a definition of microplastics, which is extensive enough to include all required criteria, has been challenging to reach.

F. Corami (✉) · B. Rosso
Institute of Polar Sciences, National Research Council – CNR ISP, Venezia, Italy

Department of Environmental Sciences, Informatics and Statistics, DAIS, University of Ca' Foscari, Venezia, Italy
e-mail: fabiana.corami@cnr.it; f_corami@unive.it

A. Khan et al. (eds.), *Microplastic sources, fate and solution*,
https://doi.org/10.1007/978-981-99-0695-6_6

Besides, sampling methods, pre-treatments, and analytical methods have not been standardized yet; hence, several methodological challenges and difficulties in comparing studies are always present [6]. Regarding size, the smaller size is strictly connected to the mesh of the sampling device (for water samples) or the technological techniques employed. As for the analytical instruments, Micro-FTIR and Raman are often employed to confirm the composition of MPs, and their limit of detection is associated with the wavelength of IR light (from 2.5 to 18 μm) [7]. Developing a specific extraction and purification procedure, the limit of detection of the smaller size can increase, i.e., for Micro-FTIR, it can be 5 μm [8].

The Technical Subgroup on Marine Litter for the implementation of the Marine Strategy Framework Directive (MSFD) recommends further subdivision of the "visible size fraction" (>1 mm) into large microplastics (1–5 mm), mesoplastics (5–25 mm), and macroplastics (>25 mm), whereas the "invisible fraction" (<1 mm) is referred to as "small microplastics" (SMPs; [7–9]). However, SMPs are considered with different upper limits by the authors, such as <500 μm [10]; between 0.1 and 1000 μm [11]; <200 μm [12]; <100 μm [8, 13, 14].

The European Chemical Agency [15] proposed a definition related to the composition of plastic polymers and their size: "MPs are solid polymer-containing particles, to which additives or other substances may have been added, with particle dimensions ranging from 1 nm to 5 mm, fiber lengths ranging from 3 nm to 15 mm, and a length-to-diameter ratio > 3". Therefore, plastic materials include an extensive range of polymer types, together with additives and plasticizers; hence, chemical identification is now an essential prerequisite to ensure the accuracy of MPs analysis [16].

The characterization of plastic polymers is essential for the correct identification of MPs in environmental matrices to improve the research reliability and discriminate the components of micro-litter from biological, organic, and other anthropogenic particles from MPs. When analyzing an environmental matrix, these latter particles can be wrongly misidentified as plastics. For instance, natural polymers (e.g., DNA, proteins, wool, silk, cellulose, cotton), paint chips, black carbon are often counted erroneously as MPs [16, 17]. The lack of polymer identification in many studies, where particles were analyzed using only microscopy, represents an obstacle to comprehensively evaluate the current number of MPs and compare the results with other studies.

Apart from size, MPs are commonly classified based on shape. MPs recorded in the research field are often pellets, fragments, and fibers, with films, ropes, filaments, sponges, foams, flakes, rubber, and microbeads [18]. However, it should be noted that different studies will use different vocabulary to categorize the same MPs. For instance, spheres, pellets, beads are often used synonymously. Besides, the words "beads" and "pellets" are strictly connected to the origin of the particles, such as microbeads in cosmetics. Some shape also represents a confusing classification, such as "fragments." It is generally used to describe particles with irregular shapes produced by environment fragmentation, which is not always the case [16]. Films are defined as thin transparency plastic, such as a piece of material from plastic bags, while foams are lighter pieces, such as foam cups [19].

Microplastic fibers are produced by fragmentation of large plastic fibers in the environment, but they directly come from washing synthetic fabrics, one of the main sources of primary MPs. They are considering one of the significant contributors to microplastic pollution in the environment, especially in urban areas. Fibrous and filamentous shapes are predominant in different matrices, such as marine and freshwater ecosystems [16]. Most of the research articles analyze long fibers ($\ll$100 μm). However, it should be highlighted that MPs are ingested according to the mouthparts' size of the organisms; hence, MP $<$100 μm are primarily ingested, and more investigations are needed on mesh sizes ($<$100 μm) [20, 21].

2 Bounty, Presence, and Conveyance of Microplastics in the Environment

2.1 Oceans Environments

The first studies of plastic debris in the North Atlantic Ocean were published at the beginning of the 1970s [22], and then an increasing number of further publications have been reported on the occurrence of plastic particles in the oceans all over the world, from North Atlantic Ocean to the Indian Ocean, even from the deepest part of the world's Mariana Trench [23]. Surface currents and winds can transport plastic litter that is fragmented into smaller pieces and accumulated to subtropical gyres, where a circular pattern of water movement and atmospheric transport allows plastics, after being transferred over long distances, to assemble in the inner [24, 25]. These five accumulation zones are distributed from the North and South Pacific Ocean, the Atlantic and South Atlantic subtropical gyre, and the Indian Ocean. The most important accumulation zone was identified and studied in the eastern part of the North Pacific Subtropical Gyre (NASG), commonly named "Great Pacific Garbage Patch." Lebreton et al. [26] predicted that at least 79 thousand tons of ocean plastic debris are floating inside this accumulation zone of 1.6 million km^2. Over three-quarters of this mass is characterized by debris larger than 5 cm, and 46% includes fishing nets. Law et al. [27] found that 83% of total plastic pieces, collected between 22° and 38°N, were equivalent to 580,000 pieces km^{-2}. A subset of samples showed that 88% were less than 10 mm in size, and most were subjected to physical degradation such as brittleness, rough edges, or fractures. Another study also found an elevated concentration of plastic particles (from 5000 to 360,000 pieces km^{-2}), and mesoplastics (5 mm–20 cm) represented about 10% of the amount of the debris collected [28]. In the garbage patch identified in the South Pacific [29], the researchers calculated an average abundance of plastic litter of 26,898 particles km^{-2}. Besides, 88.8% of plastic pollution was found near the middle of the accumulation zone. Hence, the garbage patches and floating marine litter are a relevant source of secondary MPs, even though it is difficult to establish the boundaries of these gyres accurately due to the complexity and hydrodynamics of the ocean, and consequently, of garbage patches and plastic litter.

Several point zones of accumulation have also been observed toward the Arctic Ocean [30]. Some studies [31, 32] demonstrated the importance of sea ice as a sink, source, and transport vehicle for MPs in the Arctic Ocean. Simulations and models can improve the study of MPs occurrence in the ocean as a global approach [33].

Regarding MPs pollution in the open ocean, concentrations varied from few g km^{-2} [26, 34–36] to an average abundance of 1.0×10^4 items km^{-2} in the Northwestern Pacific Ocean [37]. MPs were observed in the Eastern Indian Ocean [38], where most polypropylene (PP) and polyethylene (PE) were found (from 0.01 items m^{-2} to 4.53 items m^{-2}). The transport in open oceans has been estimated using a coupled numerical model; only ~18% of China's MPs can enter the Pacific Ocean [39]. Boundary currents along continents play a considerable role in transporting terrestrial plastic debris to the open ocean. However, it should be noted that the significant lack of knowledge on the coastal transport of MPs might contribute to a false approximation of the quantity of these pollutants, and it is unclear whether the tides also affect MPs transport in the open ocean.

MPs are characterized by having a spatial distribution, both in the water column and the sea bottom of the oceans [40]. This vertical movement may be influenced by plastic characteristics, biological processes, and different effects of hydrodynamics (i.e., internal tide movements, wind transports, Ekman pumping, upwelling, downwelling, and Stokes drift; [41]). The occurrence of MPs in the water column at different depths can vary in the several studies carried out. In the Baltic Sea water column, MPs occurrence varied from 0.07 to 2.6 items L^{-1} from the depths of 0.7–1.5 m above the bottom [42]. In the Bohai Sea, from 1.6 to 6.9 pieces L^{-1} within the size range of 100–3000 μm were present in the depths of 5–15 m [43], while from the average of 337,000 particles km^{-2} in the upper 0.5 m of the water column to 536,000 particles km^{-2} in the 5 m depth were observed [44]. Hence, the significant disparity between the expected amount and the observed MPs in the upper ocean suggests that processes can modify the distribution of MPs through the vertical water column. MPs can sink to the seafloor. Courtene-Jones et al. [45] described the widespread occurrence of MPs in sediment cores, highlighting a significant historical accumulation. MPs were found in the deep ocean on the East China Sea [46], where their abundance ranged from 0.13 to 5 items m^{-3}, with most fibers and fragments present.

Since the floating of MPs can be influenced by physical processes and biofouling, sinking MPs would be subjected to the structure of the water mass and driven by global thermohaline circulation. High-density MPs were found in the deeper layers of the water column, together with low-density plastics, suggesting that biofouling on the surface of MPs can contribute to the sinking and denaturation/degradation of plastic fragments [47, 48]. However, the vertical distribution of MPs and the mechanisms involved during the sinking process are still unknown [34, 45, 47, 49].

One of the current challenges in the MPs analysis in the ocean environments is the quantification and the chemical characterization of plastic fragments <100 μm. Small microplastics (<100 μm; SMPs) are more susceptible to vertical dispersal by mixing and dispersion and to biofouling which can decrease the resistance of plastic particles sufficiently to cause them to sink. These processes have been demonstrated to be more rapid for small particles due to their high surface-to-volume ratio [50]. SMPs were analyzed in a study at the sea surface in the North Atlantic subtropical gyre together with MPs (25–1000 μm), and SMPs are more susceptible to vertical transport [51]. Large microplastics (LMPs; 1–5 mm) were between 13,000- and 174,000-pieces km^{-2}; with a wind-driven vertical mixing correction, SMPs were estimated 300–70,000 times more abundant than LMPs. Sampling devices commonly employed, i.e., manta or neuston nets, have mesh sizes >100–300 μm [44, 52], resulting in an underestimation of SMPs. One study reported that particle numbers were 100,000 times high when employing an 80 μm mesh net instead of a 450 μm [53]. Another study analyzed MPs <333 collected in surface seawater in the North Pacific Ocean in metal buckets and immediately filtered. The authors found 8277 particles L^{-1} (8.3×10^6 particles m^{-3}) with the minimum lengths of fragments between 14 and 50 μm [54].

Quantification and characterization of MPs and SMPs comparisons in the ocean environments through different studies is difficult, especially when size distribution is taken into account [44]; further, the fate and transport of plastic in this environment remain topics for much needed further research [55].

2.2 *Marine Environments*

Microplastics are widespread in all marine environments and represent more than 95% of the general marine litters. The USA, Western Europe, Oceania, and East Asia are the most contaminated places regarded plastic pollution in marine environments. The first study of MPs in seas environments was in the 1970s, where the authors highlighted the presence of plastic litter in the Sargasso Sea surface [3], and the scientific research on the topic has been showing a heterogeneous spatial and temporal distribution of MPs. The surface, subsurface water, and sediment of different seas worldwide were investigated [56–60]. MPs in sea surface were investigated in the Slovenian part of the Trieste Bay in the Northern Adriatic, and the average abundance was $406 \cdot 10^3$ particles km^{-2} over a 20-month collecting period [61]. High concentration was also found along the Mediterranean coast of Turkey, whereas the number of floating MPs in surface water ranged from 16,339 to 520,213 items km^{-2} [62] and along the Tuscanian coast (Northwestern Mediterranean Sea) with an average of 69,161.3 items km^{-2} of floating MPs [63]. In China, the mean density of total plastic of four seasons was calculated in the surface water of the Bohai Sea with an average of $217{,}639 \pm 83{,}843$ particles km^{-2} [64]. Also, in the North Yellow Sea in China, the abundance of microplastics was high: 545 ± 282 items m^{-3} in surface seawater [65].

MPs were also investigated in sea surface waters around Scotland with a concentration ranging from 0 to 91,128 microplastics km^{-2} sea surface [66] and in remote areas, such as in Artic polar seawater with an averaged 0.34 (±0.31 SD) particles m^{-3} [67] and a median of 0.7 particles m^{-3} in subsurface waters of the Arctic Central Basin [68].

Sediments are a sink of MPs, and they are studied in different places, such as in the southern North Sea ranging between 2.8 and 1188.8 particles kg^{-1} [69], in the Southern Yellow Sea and the East China Sea off the coast of China with a mean concentration of microplastics on 13.4 ± 0.6 particles 100 g^{-1} dry weight [70] and in the Baltic sea with an average concentration of 34 ± 10 items kg^{-1} DW [71]. In this latter study, some spatial trends in MPs amounts were observed, and both fragments and films were found to sink in local sedimentation zones.

In scientific literature, much research has been reported in marine environments, but there exist a large spatial and temporal variability worldwide and the different techniques and approaches employed to identify and quantify MPs. For meaningful comparisons and monitoring, it is thus essential for the standardized protocol to correctly estimate and analyze these emerging pollutants for assessment and mitigation purposes [72].

2.3 Freshwater

MPs are identified in different freshwater ecosystems worldwide, primarily in lakes and rivers and groundwater and reservoirs, which have received less attention. It is important to underline that there is still a considerable gap between freshwater research, including sampling, pre-treatments, and techniques procedures. Hence a comparison between results from worldwide studies is difficult [73].

MPs in river environments present significant variations in abundance between different locations, and spatial, temporal, and seasonal variability is evaluated. The proximity sources in areas with a high population density and civilization can influence the evidence of MPs in freshwater environments. For example, MPs were analyzed in the Yangtze River, China, with an average of 4137 particles m^{-3}, and the most abundant plastics were fibers confirming that rivers are the crucial sources of MPs to the marine environment. The dense population concentration, river discharge, and various maritime activities influence the presence of these emerging pollutants in rivers [74]. The distance from the estuary could also influence MPs in lakes. Han et al. [75] found that MPs abundance decreased linearly with distance from the estuary in the Yellow River, China, and the average abundances of microplastics in the dry and wet seasons were 930 and 497 item L^{-1}, respectively. Further, they discovered also that the abundance of MPs in the dry season is always higher than that in the wet season. A high concentration of MPs in rivers was also found in Scotland in the River Kelvin in Glasgow. The total abundances were 161–432 MPs kg^{-1} dry sediment. As the primary type, fibers consist of >88% of total amounts and could be associated with sewage discharge. Further, the study suggests that diffuse sources of pollution may be providing secondary MPs freshwater environments [76]. In contrast, Mani et al. [77] found that 68.4% of the MPs found in the Rhine River were composed of polypropylene, whereas synthetic fibers from clothes consist primarily of polyester and acrylic fibers. The abundance of MPs analyzed was 892,777 particles km^{-2} on average, while in the Rhine-Ruhr metropolitan area, 3.9 million particles km^{-2} were evaluated. The authors also stated that MPs concentration varies across the river, suggesting different sources and sinks such as wastewater treatment plants.

Different studies were also focused on MPs in lakes. For example, many MPs were reported in Lake Geneva, Switzerland, reaching 48,146 units km^{-2} [78] or in Taihu Lake that was 3.4–25.8 items L^{-1} in surface water, and 11.0–234.6 items kg^{-1} DW in sediments [79]. Xiong et al. [80] underlined that also tourist sites could influence the occurrence of MPs in lakes; the authors found a concentration of MPs that varied between 5000 and 757,500 units km^{-2} in China's Qinghai Lake. The presence of MPs was also observed in remote areas with minimal anthropogenic impact, such as in lakeshore sediments of Tibet plateau, China, ranging from 8 ± 14 to 563 ± 1219 items m^{-2}. Morphological features suggest that MPs result from the breakdown of daily used plastic products. A massive variation of these pollutants amounts spatially in the lake, which may be associated with factors such as hydrodynamic parameters and wind [81].

Only a few studies have recently focused on other freshwater ecosystems, including natural and constructed wetlands, stormwater ponds, and reservoirs. However, there is a limited understanding of MP pollution in these compartments related to rivers, lakes, and estuaries [82]. Hence, constructed wetlands and stormwater ponds are important systems that involve stormwater and wastewater. Also, the accumulation of MPs in reservoirs is affected by human activities. Mbedzi et al. [83] studied MPs in sediment along a subtropical reservoir shoreline of Nandoni, South Africa. Liu et al. [84] found a high average amount of MPs (17,490 items kg^{-1}) in sediments from a stormwater pond in Denmark, especially in the fraction of 10–50 μm (8894 items kg^{-1}). Depending on the seasons, the amount varied between 120 and 6417 particles kg^{-1} dwt and 5–94 particles kg^{-1} dwt.

Even though the number of studies on MPs in the freshwater environment has increased over the last few years, there is heterogeneity across MP studies, and more investigations are needed regarding the effects of these pollutants in biota and humans. Further, most of the articles focused on large MPs; SMPs are then overlooked, although they are considered the most hazardous [85].

2.4 Soil

Several studies have focused on the evidence of MPs in marine environments, and knowledge on MPs pollution in terrestrial environments is relatively restricted [86]. It has been estimated that soils may contain more plastic waste pollution than the environments of the oceans [87]. MPs can move into soil environments through different pathways, such as land application of sludge and organic fertilizer, plastic mulching, atmospheric deposition or wastewater irrigation, and surface runoff [86].

The proximity of anthropogenic activities can influence the MPs' evidence in soil. For instance, soils in Australia near an industrial area contain a range of 300–67,500 mg kg^{-1} of MPs [88]. Industrial production processes are confirmed to increase MP abundance in the soil in China. Zhou et al. [89] found that the highest total MP concentration was in an industrial area with a close chemical fiber factory. MPs are also transferred into the soil due to sewage sludge for agricultural applications. In China, lands used for plastic mulching (mulch film and greenhouse crops) contain significantly high MPs in the range of 2783–6366 items kg^{-1} in all samples [90]. Agricultural soils have also been investigated in the lower reaches of the Yangtze River, China, with an average of 37.32 items kg^{-1} of MPs abundance [91]. A similar result was obtained in the Qinghai-Tibet plateau, where the primary MP source was the application of plastic mulch in farmlands, and the average concentration of MPs was between 53.2 ± 29.7 and 43.9 ± 22.3 items kg^{-1} [14], while in farmland soils in suburbs of Shanghai, China, an average abundance of MPs was 78.00 and 62.50 items kg^{-1} in the top and deep soils [92]. However, a low abundance of MPs was observed in agricultural farmland in southeast Germany with 0.34 ± 0.36 MPs particles kg^{-1} dry weight of soil [93].

This study highlights the impact of artificial activities on soil microplastics in coastal areas, for instance, mariculture, tourism, and port construction in coastal soils adjacent to the Bohai Sea and the Yellow Sea [94], ranging from 1.3 to 14,712.5 N kg^{-1} (dry weight). Low abundances of microplastics are reported in floodplain areas of the Lahn River, Germany ranged between 2.06 kg^{-1} (±1.55 kg^{-1}) and 1.88 kg^{-1} (±1.49 kg^{-1}) [95] and in the floodplain soils in Switzerland (55.5 mg kg^{-1} and 593 particles kg^{-1}) [96].

2.5 *Atmospheric Compartment*

While MPs have been studied in many environmental matrices, their occurrence in the atmospheric compartment was only studied recently. Concentrations differ significantly among research due to methodologies and sampling procedures. MP evidence is confirmed in the atmospheric compartment, but knowledge on the depositions, transport, and impacts is limited. The results of atmospheric fallout are presented as deposition rates (MP number m^{-2} day^{-1}) in literature; in the case of suspended particles, the results are shown as a concentration in MP number m^{-3} [97].

2.5.1 Urban Areas

The first study was published by Dris et al. [98], which found in total atmospheric fallout 29–280 particles m^{-2} day^{-1} in the urban area of Paris, France. The range of the resulting MPs analyzed was from 100 to 5000-μm size, and fibers were the most common. Then, the study was amplified to investigate the atmospheric fallout between urban and suburban sites in Paris. The atmospheric fallout ranged from 2 to 355 particles m^{-2} day^{-1}, with an average of 110 ± 96 particles m^{-2} day^{-1} in the urban site and a lower amount of 53 ± 38 particles m^{-2} day^{-1} in the suburban site due to the lower density of the surrounding population. The smallest fibers (in the 200–400 μm and 400–600 μm size ranges) are the major. Fibers smaller than <50 μm are also observed, but they are hard to identify; hence, they were not taken into account. Fibers were principally made of polyethylene terephthalate (PET) and polyethylene (PE). This study showed an estimate between 3 and 10 tons of fibers deposited by atmospheric fallout every year in Paris, underlined the importance of MPs in the atmospheric compartment, especially in urban areas [99]. Another study in a big urban area was performed by Cai et al. [100] in Dongguan city, China, and the MPs concentrations range of 175–313 particles m^{-2} day^{-1} in the atmospheric fallout. Also, in this study, fibers were dominant, and PE, PP, and polystyrene (PS) were identified. The abundance of MPs in the atmospheric fallout from Dongguan city is significantly closer to the urban site in Paris and has the same order of magnitude overall. A lower concentration was found by Liu et al. [101] in Shanghai, China, and suspended atmospheric MPs ranged from 0 to 4.18 n m^{-3}. The vertical distribution of atmospheric MPs was evaluated, and their abundances tended to decrease horizontally. The highest amount was detected at 1.7 m on the ground, reducing abundance with a higher altitude. 67% of all MPs observed were fibers; particles in this main shape ranged in size from 23.07 to 9555 μm, with an average of 582.2 μm. Composition analysis showed that different plastic composition was investigated, especially PES (polyester). Also, in this study, a first total estimation of the annual weight of suspended atmospheric MPs was evaluated in Shanghai, which is nearly 120.72 kg.

Similar low concentrations were observed by Abbasi et al. [102] in Asaluyeh County, Iran-industrial station. The total suspended particulates were about 1 m^{-3} ranging from about 2 to 5000 μm with the ubiquity of MPs fibers. More than 75% of particles occur in the <100 μm category, and gradually smaller proportions are present with rising size.

2.5.2 Outdoor and Indoor Area

Considering indoor and outdoor air is important to understand the microplastic dynamics in an urban environment and the impact on biota and human health.

One of the first indoor and outdoor air studies was evaluated for airborne fibers by Dris et al. [103] in France. Indoor concentrations (apartments and office) ranged between 1.0 and 60.0 fibers m^{-3}, while outdoor are considerably lower as they range between 0.3 and 1.5 fibers m^{-3}. 67% of the analyzed fibers in indoor environments are made of principal cellulose, while the remaining contain mainly PP. Regarding the size, the range of 4650–4850 μm was analyzed in dustfall, and no fiber longer than 3250 μm was detected in indoor air. The size of the fibers in outdoor air is constantly smaller than 1650 μm. It is noticed that fibers less than 50 μm were not counted due to the observation lower limit. Also, in Aveiro, Portugal, a study [10] revealed concentrations of 6 fiber m^{-3} for indoor and outdoor, while the suspected microplastics were higher outdoor (12 MPs m^{-3}) than indoor, where 5 particles m^{-3} were found. Regarding the variation, fiber size ranges from 17 to 3669 μm in indoor samples and 26 to 1928 μm in outdoor samples. However, the nature of the identified fibers was unidentified (62.3%) due to their small sizes.

A higher concentration of MP abundance was found in indoor (1583 ± 1180 n m^{-3}) and outdoor (189 ± 85 n m^{-3}) air from urban areas of a coastal city in eastern China [104]. In contrast with other studies, SMPs were dominant (>90%) in indoor and outdoor air samples in this work. The most common polymer composition among indoor was PES, PA, and PP, while PE, PS, and PES dominated outdoor samples. Regarding the shape, fragments were dominant compared to fibers in all samples. Vianello et al. [105] found PES and PE MPs, together with nylon, in indoor airborne using a Breathing Thermal Manikin in Denmark. The concentration was between 1.7 and 16.2 particles m^{-3}. Synthetic fragments and fibers reported were 4% of the total particles, while nonsynthetic particles of protein and cellulose constituted the majority.

2.5.3 Rural Versus Urban Area and Remote Areas

Despite the lack of clear evidence, the authors suspect that MPs are more concentrated in high human activity areas [97], but some studies revealed a higher concentration in remoteness regions. For example, a comparison between atmospheric fallout in urban and rural environments was investigated by Klein and Fischer [106] in the region of Hamburg, Germany. Unexpectedly, they found a high particle concentration in the rural sampling sites, and fragments were substantially dominating compared to fibers. The authors suspected that the MPs amount are affected by local effects, such as the closeness of a highway and agricultural activities. The concentration of MPs was given in particles m^{-2} day^{-1}: in the urban area ranged from 136.5 to 260.6 particles m^{-2} day^{-1}, while in rural areas from 331.4 to 512.0 particles m^{-2} day^{-1}. Polyethylenes/ethyl vinyl acetate copolymers were the most abundant. The atmosphere compartment is an essential contributor to the transport of MPs in all environments, and the air–sea interactions controlled by the cloud cover together with meteorological parameters play a crucial role in the worldwide transport of atmospheric particles [107]. Liu et al. [108] studied the atmospheric transport of suspended atmospheric MPs from terrestrial to the marine compartment through a cruise in the western Pacific Ocean. A high abundance of MPs was observed in the atmosphere along the coasts (0.13 ± 0.24 items m^{-3}). In comparison, less quantity was detected in the pelagic area (0.01 ± 0.01 items m^{-3}), which proved significant atmospheric transport of MPs to the marine environment, especially the pollution caused by textile microfibers. Regarding the size, MPs ranged from 16.14 to 2086.69 μm, with an average of 318.53 μm, and PET particles were the most common composition analyzed.

Studies have many differences from the scientific literature results in the atmospheric compartment, mainly due to different methodologies connected to detecting particles at a different cut-off size, using different samplers and detection methodology techniques [106]. The smallest class of particles is frequently underrepresented because of identification artifacts, but SMPs need to be investigated because they represent a significant health concern [97]. Besides, the current research was only studied in a few cities or regions with a limited sampling period. Spatial and temporal variations of airborne atmospherics and transport in the environments need to be further investigated together with standardized operation protocols to reproduce the results worldwide [109].

2.6 Biota

The awareness of detecting MPs in biotic compartments of different environments has increased, with numerous in-depth studies on MPs' physical-biological and chemical effects in different organisms from low and high trophic levels. The first impact of plastic waste in biota was reported in the 1960s, with a study of plastic particles in the gastrointestinal system of birds in the marine environment [110]. Then, MPs have been investigated in all environmental compartments, even in remote places [111–113], and the entrance into the food web of these emerging pollutants, especially of SMPs, represents a significant concern.

MP research on biota has focused on higher taxa, specifically fish and mammals, with fewer environmental studies assessing lower taxa. It is estimated that 60% globally of fish contained MPs, but most studies focused on gastrointestinal tracts (GITs), with presence in other organs and tissues currently being too little evaluated. Fish are primarily studied in marine environments (52%), flowed by freshwater (31%), mangrove, and estuarine [114]. GIT was investigated in 100 individuals from 10 commercial marine fish species from the Bay of Bengal with an average abundance in the range of 2.2 ± 0.89 MPs per individual [115]. A similar range was analyzed in GIT of 14 fish species from the Persian Gulf with 2.46 ± 1.46 MPs per individual [116]; in 7 commercially targeted fish species from the Agulhas Bank, south of South Africa that MPs ranged from 2.8 to 4.6 items per fish [117] and in fish from the western Mediterranean Sea with a mean value of 1.13 ± 0.16 particles per individual [118]. Higher results were obtained by analyzing fish gut from an important fish farm and mariculture area, Haizhou Bay, China, and the highest abundance of MPs was 22.21 ± 1.70 items per individual [119]. However, a review on MPs in fish and fishmeal reported that 55% of the research on this topic was unsuitable due to the focus on large microplastics (>1 mm) with a lack of or limited contamination control and polymer characterization [120].

Marine mammals and sea turtles are considered long-lived sentinel species that are frequently used for biomonitoring the health status of the ocean. For instance, the average number of MPs found in the GI of large odontocetes ranged from 9 to 88 MPs per individual, with a size ranging from 0.1 to 5 mm [121]. According to a review [122], 88% of turtles evaluated are polluted by MPs, with an average of 121 items per individual. Turtles are the group most affected by MPs, followed by 59% of marine mammals and 50% of sea birds. This review also reports that the average MPs particles in marine mammals are 9.7 particles for individuals and 7.0 for sea birds. 25% of the articles described analyzing MPs between 0.3 and 0.5 mm, while the rest focused on a size >0.5 mm, especially fibers.

Invertebrates are a heterogeneous group including different taxa. According to European Water Framework Directive 2000/60/EC, invertebrates are significant bioindicators for water quality and their ecological role because they are at the base of trophic webs. A more comprehensive understanding of the ingestion of microplastics by riverine macroinvertebrates is needed given their frequent position as primary consumers supporting riverine food webs and their potential use for determining the origins and entry points of MPs in freshwater food webs.

3 Danger and Impacts of Microplastics

3.1 Toxicological Impacts

The characteristics of resistance, durability, and persistence in the environments make MPs a potential threat to biota and human health worldwide. Hence, MPs are considering a vector in the environment of toxic chemical contaminants adsorbed to the MPs surface. Due to their relatively large specific surface area, MPs can adsorb organic and inorganic chemical contaminants, e.g., trace elements, persistent organic pollutants (POPs) such as polyaromatic hydrocarbons (PAHs), organochlorine pesticides (OCPs), and polychlorinated biphenyls (PCBs). Besides, polymers contain plasticizers (i.e., phthalates), brominated flame retardants, antioxidants, processing chemicals, and pigments and colors [123, 124].

The presence of MPs can induce different toxicological effects, such as oxidative stress; altered gene expressions; disrupted endocrine system; neurotoxicity; immune responses, reproductive abnormities, and death [125–129]. For example, recent research on fish early life stages behavior discovered that survival, hatching rates, and swimming activity of zebrafish larvae decreased after prolonged exposure to MPs and Cu [130]. Toxicological effects were also studied in terrestrial organisms, such as insects or earthworms. For example, molecular effects of tire rubber and PS were evaluated in *Chironomus riparius*, and they found that overexpression of genes was related to the development and oxidative stress. Further, these effects occurred mainly in response to tire micro rubber [125]. The effects of low concentrations of PS particles (100 and 1300 nm) were also studied in earthworms, and they found that the presence of these particles can induce oxidative stress, histopathological changes, and DNA damage can happen to earthworms [131].

Specific properties of MPs can influence toxicological effects in biota. Hence, polymer size, shape, density, and types can impact sorption dynamics [132]. Kim and An [133] discovered that the average velocity and distance movement of springtail *Folsomia candida* decreases in response to MPs ingestion, highlighting the harmful effects of MPs on soil biota. The results highlighted that MPs smaller than the edible size of 66.0 ± 10.9 μm could substantially reduce these movements in springtails. Also, glass transition temperature, degree of cross-linking, and crystallinity are parameters that can influence the sorption behavior between MPs and pollutants [134]. For example, the glass transition temperature of MPs and the development of a hydrophilic surface layer after the aging process were investigated by Muller et al. [135]. PP showed a higher sorption capacity for the fuel-related water contaminants than PS, while glass transition temperature principally affected PP and PS's sorption behavior. Aging effects do not influence PP's sorption, while PS sorbed fewer pollutants due to the presence of an oxidized surface layer.

Another toxicological aspect of the presence of MPs in the environments is correlated to the sorption of microbial and pathogens. The dominance of bacterial species may contribute to biofilm formation that enables consequent colonization and proliferation of other pathogens, increasing their hazards in all environmental matrices. Pham et al. [136] showed that in municipal activated sludge, bacteria and pathogens on MPs could become antibiotic-resistant (especially Novosphingobium and Flectobacillus). Besides, MPs can also accumulate fungal pathogens in terrestrial ecosystems. Gkoutselis et al. [137] showed that MPs act as specific artificial microhabitats that attract distinct fungal communities and opportunistic human pathogens, such as cryptococcal and Phoma-like species. These results highlighted the potentially severe consequences for the trans-kingdom and different biota epidemiology of fungal infections in terrestrial environments worldwide.

3.2 Bioaccumulation Through the Food Net

MPs can be ingested and accumulated by lower trophic levels biota with potential bioaccumulation and biomagnification into the trophic web, either through direct ingestion or indirectly by trophic transfer from contaminated prey. However, only a few studies focus on the bioaccumulation of MPs, and more investigations are needed. MPs have been found inside the stomach and gastrointestinal tracts (GIT) of different organisms, such as various commercial and wild fish species [115, 138], highlighting the potential for transfer to predators to occur. The consumption of polluted prey can also affect marine mammals and birds. For example, it is calculated that a striped dolphin (*Stenella coeruleoalba*) can ingest about 463 million MPs by one through the consumption of contaminated prey [139]. 100% of little auks, seabirds from the Greenland sea, can ingest often more light-colored MPs rather than darker ones, strongly suggesting active contamination with birds mistaking MPs for their natural zooplankton prey [140].

Besides the bioaccumulation of MPs in the digestive tracts, different studies have demonstrated that MPs can be translocated to accumulate in tissues. MP biomagnification through the marine trophic web has been explored in a variety of commercial species such as in crabs through mussel ingestion, in shrimps from zooplankton, and in fur seals from pelagic fish; but harmonized protocols including techniques and methods for MP analysis in environmental matrixes and how the exposure to MPs poses a risk for human health are needed to evaluate in future [141].

The size of MPs is a significant factor to consider focused on bioaccumulation through the trophic web. Hence, most scientific reports focused on assessing accumulated MPs >1 mm in the gut of different species after ingestion with the possibility to expel via feces, starvation, or suffocation consequences. [142]. Fernandez and Almendosa [143] showed that mussels exposed to MPs (HDPE, ≤10 μm) and left to depurate for 7 days. While SMPs could still be eliminated, MPs may accumulate in the digestive gland and the gills, decreasing with time but increasing MPs' burdens during depuration. However, SMPs could be accumulated as well; the harmful persistence of ≤4 μm size MPs underlined the massive problem of organisms in the marine environment. Initial experiments showed that upon ingestion, microplastic accumulated in the gut. The fish *Danio rerio* can recognize MPs as inedible materials, but MPs are ingested when their size ranges between 1 and 5 μm; these particles can be accumulated in the digestive tract after one or multiple meals. Although this species can almost totally evacuate MPs after 24 h, few particles remain in the digestive tract [144].

The trophic transfer is an indirect pathway of MPs ingestion for different species whose feeding affects prey consumption, including humans. Commercial seafood is one of the significant pathways of MPs regarding human health, and the bioaccumulation and biomagnification of MPs in the trophic web may pose a severe threat to food safety [141]. Only recently, MPs have been found in the human body. Ragusa et al. [145] found 12 microplastic fragments in 4 different human placentae; or in human lung tissues where PE and PP fragments were >5.5 μm in size [146].

The accumulation of microplastic particles in humans has potential health risks such as cytotoxicity, hypersensitivity, unwanted immune response, and acute response like hemolysis [147]. In human bodies, PP particles (in different concentrations and sizes of approximately ~20 μm and 25–200 μm) can cause health problems by inducing the production of cytokines from immune cells rather than by direct toxicity. The authors also underline that there are also thousands of other types of plastic in different concentrations and sizes in the environment that should be studied and need more investigations in the future.

References

1. Plastics Europe (2019) Plastics—The Facts 2019: an analysis of european plastics production, demand and waste data. PlasticsEurope
2. Geyer R, Jambeck JR, Law KL (2017) Production, use, and fate of all plastics ever made. Sci Adv 3:7. https://doi.org/10.1126/sciadv.1700782
3. Carpenter EJ, Smith KL (1972) Plastics on the Sargasso sea surface. Science 175:1240–1241. https://doi.org/10.1126/science.175.4027.1240
4. Morris RJ (1980) Floating plastic debris in the Mediterranean. Mar Pollut Bull 11(5):125. https://doi.org/10.1016/0025-326X(80)90073-9
5. Thompson R, Olsen Y, Mitchell RP, Davis A, Rowland JS, John AWG, Mcgonigle D, Russell AE (2004) Lost at sea: where is all the plastic? Science 304(5672):838
6. Frias JPGL, Nash R (2019) Microplastics: finding a consensus on the definition. Mar Pollut Bull 138:145–147. https://doi.org/10.1016/j.marpolbul.2018.11.022. Epub 2018 Nov 22

7. Filella M (2015) Questions of size and numbers in environmental research on microplastics: methodological and conceptual aspects. Environ Chem 12:527–538. https://doi.org/10.1071/EN15012. https://archive-ouverte.unige.ch/unige:74828
8. Corami F, Rosso B, Roman M, Picone M, Gambaro A, Barbante C (2020) Evidence of small microplastics (<100 μm) ingestion by Pacific oysters (Crassostrea gigas): a novel method of extraction, purification, and analysis using Micro-FTIR. Mar Pollut Bull 160:111606. https://doi.org/10.1016/j.marpolbul.2020.111606
9. Galgani F, Hanke G, Werner SDVL, De Vrees L (2013) Marine litter within the European marine strategy framework directive. ICES J Mar Sci 70(6):1055–1064. https://doi.org/10.1093/icesjms/fst122
10. Prata JC, Castro JL, da Costa JP, Duarte AC, Rocha-Santos T, Cerqueira M (2020) The importance of contamination control in airborne fibers and microplastic sampling: experiences from indoor and outdoor air sampling in Aveiro, Portugal. Mar Pollut Bull 159:111522. https://doi.org/10.1016/j.marpolbul.2020.111522
11. Rani-Borges B, Moschini-Carlos V, Pompêo M (2021) Microplastics and freshwater microalgae: what do we know so far? Aquat Ecol 55:363–377. https://doi.org/10.1007/s10452-021-09834-9
12. Zhou Y, He G, Jiang X, Yao L, Ouyang L, Liu X et al (2021) Microplastic contamination is ubiquitous in riparian soils and strongly related to elevation, precipitation and population density. J Hazard Mater 411:125178. https://doi.org/10.1016/j.jhazmat.2021.125178
13. Corami F, Rosso B, Morabito E, Rensi V, Gambaro A, Barbante C (2021) Small microplastics (<100 μm), plasticizers and additives in seawater and sediments: oleo-extraction, purification, quantification, and polymer characterization using Micro-FTIR. Sci Total Environ 797:148937. https://doi.org/10.1016/j.scitotenv.2021.148937
14. Feng S, Lu H, Liu Y (2021) The occurrence of microplastics in farmland and grassland soils in the Qinghai-Tibet plateau: different land use and mulching time in facility agriculture. Environ Pollut 279:116939. https://doi.org/10.1016/j.envpol.2021.116939
15. ECHA (European Chemicals Agency) (2019) Annex XV restriction report proposal for a restriction. Report Version Number 1 (March 20th, 2019). (Helsinki)
16. Hartmann NB, Hüffer T, Thompson RC, Hassellöv M, Verschoor A, Daugaard AE, Rist S, Karlsson T, Brennholt N, Cole M, Herrling MP, Hess MC, Ivleva NP, Lusher AL, Wagner M (2019) Environ Sci Technol 53(3):1039–1047
17. Ivar do Sul JA (2021) Why it is important to analyze the chemical composition of microplastics in environmental samples. Mar Pollut Bull 165:112086
18. Rosal R (2021) Morphological description of microplastic particles for environmental fate studies. Mar Pollut Bull 171:112716. https://doi.org/10.1016/j.marpolbul.2021.112716
19. Acharya S, Rumi SS, Hu Y, Abidi N (2021) Microfibers from synthetic textiles as a major source of microplastics in the environment. Text Res J 91(17-18):2136–2156
20. Rebelein A, Int-Veen I, Kammann U, Scharsack JP (2021) Microplastic fibers – underestimated threat to aquatic organisms? Sci Total Environ 777:146045
21. Walkinshaw C, Lindeque PK, Thompson R, Tolhurst T, Cole M (2020) Microplastics and seafood: lower trophic organisms at highest risk of contamination. Ecotoxicol Environ Saf 190110066
22. Wong CS, Green D, Cretney WJ (1974) Quantitative tar and plastic waste distributions in the Pacific Ocean. Science 247:30–32
23. Peng X, Chen M, Chen S, Dasgupta S, Xu H, Ta K, Du M, Li J, Guo Z, Bai S (2018) Microplastics contaminate the deepest part of the world's ocean. Geochem Perspect Lett 9. https://doi.org/10.7185/geochemlet.1829
24. Eriksen M, Lebreton LCM, Carson HS, Thiel M, Moore CJ, Borerro JC, Galgani F, Ryan PG, Reisser J (2014) Plastic pollution in the world's oceans: more than 5 trillion plastic pieces weighing over 250,000 tons afloat at sea. PLoS One 9(12):e111913. https://doi.org/10.1371/journal.pone.0111913

25. Moore CJ (2008) Synthetic polymers in the marine environment: a rapidly increasing, long-term threat. Environ Res 108:131–139
26. Lebreton L, Slat B, Ferrari F, Sainte-Rose B, Aitken J, Marthouse R, Hajbane S, Cunsolo S, Schwarz A, Levivier A, Noble K, Debeljak P, Maral H, Schoeneich-Argent R, Brambini R, Reisser J (2018) Evidence that the Great Pacific Garbage Patch is rapidly accumulating plastic. Sci Rep 8:4666. https://doi.org/10.1038/s41598-018-22939-w
27. Law KL, Morét-Ferguson K, Maximenko NA, Proskurowski G, Peacock EE, Hafner J, Reddy CM (2010) Plastic accumulation in the North Atlantic subtropical gyre. Science 329(5996): 1185–1188. https://doi.org/10.1126/science.1192321
28. Brach L, Deixonne P, Bernard M, Durand E, Desjean M, Perez E, van Sebille E, ter Halle A (2018) Anticyclonic eddies increase accumulation of microplastic in the North Atlantic subtropical gyre. Mar Pollut Bull 126:191–196. https://doi.org/10.1016/j.marpolbul.2017.10.077
29. Eriksen M, Mason S, Wilson S, Box C, Zellers A, Edwards W, Farley H, Amato S (2013) Microplastic pollution in the surface waters of the Laurentian Great Lakes. Mar Pollut Bull 77: 177–182. https://doi.org/10.1016/j.marpolbul.2013.10.007
30. Lima ARA, Ferreira GVB, Barrows APW, Christiansen KS, Treinish G, Toshack MC (2021) Global patterns for the spatial distribution of floating microfibers: Arctic Ocean as a potential accumulation zone. J Hazard Mater 403:123796. https://doi.org/10.1016/j.jhazmat.2020.123796
31. Kanhai LDK, Gardfeldt K, Krumpen T, Krumpen T, Thompson RC, O'Connor I (2020) Microplastics in sea ice and seawater beneath ice floes from the Arctic Ocean. Sci Rep 10: 5004. https://doi.org/10.1038/s41598-020-61948-6
32. Peeken PS, Beyer B, Gütermann J, Katlein C, Krumpen T, Bergmann M, Hehemann L, Gerdts G (2018) Arctic sea ice is an important temporal sink and means of transport for microplastic. Nat Commun 9:1505. https://doi.org/10.1038/s41467-018-03825-5
33. Khatmullina L, Chubarenko I (2019) Transport of marine microplastic particles: why is it so difficult to predict? Anthropocene Coasts 2. https://doi.org/10.1139/anc-2018-0024
34. Cózar A, Echevarría F, González-Gordillo JI, Irigoien X, Úbeda B, Hernández-León S, Palma AT, Navarro S, García-de-Lomas J, Ruiz A, Fernández-de-Puelles ML, Duarte CM (2014) Plastic debris in the open ocean. PNAS 111(28):10239–10244. https://doi.org/10.1073/pnas.1314705111
35. Goldstein MC, Rosenberg M, Cheng L (2012) Increased oceanic microplastic debris enhances oviposition in an endemic pelagic insect. Biol Lett 8:817–820. https://doi.org/10.1098/rsbl.2012.0298
36. Law KL, Thompson RC (2014) Microplastics in the seas. Science 345(6193):144–145. https://doi.org/10.1126/science.1254065
37. Pan Z, Guo H, Chen H, Wang S, Sun X, Zou Q, Zhang Y, Lin H, Cai S, Huang J (2019) Microplastics in the Northwestern Pacific: abundance, distribution, and characteristics. Sci Total Environ 650(2):1913–1922. https://doi.org/10.1016/j.scitotenv.2018.09.244
38. Li C, Wang X, Liu K, Zhu L, Wei N, Zong C, Li S (2021) Pelagic microplastics in surface water of the Eastern Indian Ocean during monsoon transition period: abundance, distribution, and characteristics. Sci Total Environ 755(2):142629. https://doi.org/10.1016/j.scitotenv.2020.142629
39. Zhang Z, Wu H, Peng G, Xu P, Li D (2020) Coastal ocean dynamics reduce the export of microplastics to the open ocean. Sci Total Environ 713:136634. https://doi.org/10.1016/j.scitotenv.2020.136634
40. Woodall LC, Sanchez-Vidal A, Canals M, Paterson GLJ, Coppock R, Sleight V, Calafat A, Rogers AD, Narayanaswamy BE, Thompson RC (2014) The deep sea is a major sink for microplastic debris. R Soc Open Sci 1(4):140317. https://doi.org/10.1098/rsos.140317
41. van Sebille E, Aliani S, Law KL, Maximenko N, Alsina JM, Bagaev A, Bergmann A, Chapron A, Chubarenko I, Cózar A, Delandmeter P, Egger P, Fox-Kemper P, Garaba SP, Goddijn-Murphy L, Hardesty BD, Hoffman MJ, Isobe A, Jongedijk CE, Kaandorp MLA,

Khatmullina L, Koelmans ukulka T, Laufkötter C, Lebreton L, Lobelle D, Maes C, Martinez-Vicente V, Morales Maqueda MA, Poulain-Zarcos M, Rodríguez E, Ryan PG, Shanks AL, Shim WJ, Suaria G, Thiel M, van den Bremer TS, Wichmann D (2020) The physical oceanography of the transport of floating marine debris. Environ Res Lett 15(2):023003. https://doi.org/10.1088/1748-9326/ab6d7d
42. Bagaev A, Mizyuk A, Khatmullina L, Isachenko I, Chubarenko I (2017) Anthropogenic fibres in the Baltic Sea water column: field data, laboratory and numerical testing of their motion. Sci Total Environ 599-600:560–571. https://doi.org/10.1016/j.scitotenv.2017.04.185
43. Dai Z, Zhang H, Zhou Q, Tian Y, Chen T, Tu C, Fu C, Luo Y (2018) Occurrence of microplastics in the water column and sediment in an inland sea affected by intensive anthropogenic activities. Environ Pollut 242:1557–1565. https://doi.org/10.1016/j.envpol.2018.07.131
44. Kooi M, Reisser J, Slat B, Ferrari FF, Schmid MS, Cunsolo S, Brambini R, Noble K, Sirks L, Linders TEW, Schoeneich-Argent RI, Koelmans AA (2016) The effect of particle properties on the depth profile of buoyant plastics in the ocean. Sci Rep-Uk 6:1. https://doi.org/10.1038/srep33882
45. Courtene-Jones W, Quinn B, Ewins C, Gary SF, Narayanaswamy BE (2020) Microplastic accumulation in deep-sea sediments from the Rockall Trough. Mar Pollut Bull 154:111092. https://doi.org/10.1016/j.marpolbul.2020.111092
46. Liu K, Zhang F, Song Z, Zong C, Wei N, Li D (2019) A novel method enabling the accurate quantification of microplastics in the water column of Deep Ocean. Mar Pollut Bull 146:462–465. https://doi.org/10.1016/j.marpolbul.2019.07.008
47. Liu K, Courtene-Jones W, Wang X, Song Z, Wei N, Li D (2020) Elucidating the vertical transport of microplastics in the water column: a review of sampling methodologies and distributions. Water Res 186:116403. https://doi.org/10.1016/j.watres.2020.116403
48. Kaiser D, Kowalski N, Waniek JJ (2017) Effects of biofouling on the sinking behavior of microplastics. Environ Res Lett 12:124003. https://doi.org/10.1088/1748-9326/aa8e8b
49. Kvale K, Prowe AEF, Chien CT, Landolfi A, Oschlies A (2020) The global biological microplastic particle sink. Sci Rep 10:16670. https://doi.org/10.1038/s41598-020-72898-4
50. Pabortsava K, Lampitt RS (2020) High concentrations of plastic hidden beneath the surface of the Atlantic Ocean. Nat Commun 11:4073. https://doi.org/10.1038/s41467-020-17932-9
51. Poulain M, Mercier MJ, Brach L, Martigna M, Routaboul C, Perez E, Desjean MG, ter Halle A (2019) Small microplastics as a main contributor to plastic mass balance in the North Atlantic subtropical gyre. Environ Sci Technol 53(3):1157–1164. https://doi.org/10.1021/acs.est.8b05458
52. Eriksen M, Liboiron M, Kiessling T, Charron L, Alling A, Lebreton L, Richards H, Roth B, Ory NC, Hidalgo-Ruz V, Meerhoff E, Box C, Cummins A, Thiel M (2018) Microplastic sampling with the AVANI trawl compared to two neuston trawls in the Bay of Bengal and South Pacific. Environ Pollut 232:430–439. https://doi.org/10.1016/j.envpol.2017.09.058
53. Norén F (2007) Small plastic particles in Coastal Swedish waters. Tech Rep 0, KIMO. https://www.researchgate.net/profile/Fredrik-Noren/publication/284312290_Small_plastic_particles_in_Coastal_Swedish_waters/links/571203c608ae4ef74525ec38/Small-plastic-particles-in-Coastal-Swedish-waters.pdf
54. Brandon JA, Freibott A, Sala LM (2020) Patterns of suspended and salp-ingested microplastic debris in the North Pacific investigated with epifluorescence microscopy. Limnol Oceanogr 5:46–53. https://doi.org/10.1002/lol2.10127
55. Galgani F, Brien AS, Weis J, Ioakeimidis C, Schuyler Q, Makarenko I, Griffiths H, Bondareff J, Vethaak D, Deidun A, Sobral P, Topouzelis K, Vlahos P, Lana F, Hassellov M, Gerigny O, Arsonina B, Ambulkar A, Azzaro M, Bebianno MJ (2021) Are litter, plastic and microplastic quantities increasing in the ocean? Micropl Nanopl 1:2. https://doi.org/10.1186/s43591-020-00002-8

56. Nunes BZ, Huang Y, Ribeiro VV, Wu S, Holbech H, Moreira LB, Xu EG, Castro IB (2022) Microplastic contamination in seawater across global marine protected areas boundaries. Environ Pollut 316(Pt 1):120692. https://doi.org/10.1016/j.envpol.2022.120692
57. Yin Z (2022) The pollution of microplastics in sediments: the ecological risk assessment and pollution source analysis. Sci Total Environ 859(Part 2):160323. https://doi.org/10.1016/j.scitotenv.2022.160323
58. Andrady AL (2011) Microplastics in the marine environment. Mar Pollut Bull 62:1596–1605. https://doi.org/10.1016/j.marpolbul.2011.05.030
59. Kukulka AA, Laufkötter T, Lebreton C, Lobelle L, Maes D, Martinez-Vicente C, Morales M, Maqueda M, Poulain-Zarcos M, Rodríguez R, Ryan PG, Shanks AL, Shim WJ, Suaria G, Thiel M, van den Bremer TS, Wichmann D (2020) The physical oceanography of the transport of floating marine debris. Environ Res Lett 15(2):23003–23032
60. Maximenko N et al (2019) Toward the integrated marine debris observing system. Front Mar Sci 6:447. https://www.frontiersin.org/article/10.3389/fmars.2019.00447
61. Gajšt T, Bizjak T, Palatinus A, Liubartseva S, Kržan A (2016) Sea surface microplastics in Slovenian part of the northern Adriatic. Mar Pollut Bull 113:392–399. https://doi.org/10.1016/j.marpolbul.2016.10.031
62. Güven O, Gökdağ K, Jovanović B, Kıdeyş AE (2017) Microplastic litter composition of the Turkish territorial waters of the Mediterranean Sea, and its occurrence in the gastrointestinal tract of fish. Environ Pollut 223:286–294. https://doi.org/10.1016/j.envpol.2017.01.025
63. Baini M, Fossi MC, Galli M, Caliani I, Campania T, Finoi MG, Panti C (2018) Abundance and characterization of microplastics in the coastal waters of Tuscany (Italy): the application of the MSFD monitoring protocol in the Mediterranean Sea. Mar Pollut Bull 133:543–552. https://doi.org/10.1016/j.marpolbul.2018.06.016
64. Zhang W, Zhang S, Zhao Q, Qu L, Ma D, Wang J (2020) Spatio-temporal distribution of plastic and microplastic debris in the surface water of the Bohai Sea, China. Mar Pollut Bull 158:111343. https://doi.org/10.1016/j.marpolbul.2020.111343
65. Zhu L, Bai Z, Chen B, Sun X, Qu K, Xi B (2018) Microplastic pollution in North Yellow Sea, China: observations on occurrence, distribution and identification. Sci Total Environ 636:20–29. https://doi.org/10.1016/j.scitotenv.2018.04.182
66. Russell M, Webster L (2021) Microplastics in sea surface waters around Scotland. Mar Pollut Bull 166:112210. https://doi.org/10.1016/j.marpolbul.2021.112210
67. Lusher AL, Tirelli V, O'Connor I, Officer R (2015) Microplastics in Arctic polar waters: the first reported values of particles in surface and sub-surface samples. Sci Rep 5:14947. https://doi.org/10.1038/srep14947
68. Kanhai DK, Gårdfeldt K, Lyashevska O, Hassellöv M, Thompson RC, O'Connor I (2018) Microplastics in sub-surface waters of the Arctic Central Basin. Mar Pollut Bull 130:8–18. https://doi.org/10.1016/j.marpolbul.2018.03.011
69. Lorenz C, Roscher L, Meyer MS, Hildebrandt L, Prume J, Löder MJG, Primpke S, Gerdts G (2019) Spatial distribution of microplastics in sediments and surface waters of the southern North Sea. Environ Pollut 252, B:1719–1729. https://doi.org/10.1016/j.envpol.2019.06.093
70. Zhang C, Zhou H, Cui Y, Wang C, Li Y, Zhang D (2019) Microplastics in offshore sediment in the Yellow Sea and East China Sea, China. Mar Pollut Bull 114(2):724–732. https://doi.org/10.1016/j.envpol.2018.10.102
71. Zobkov M, Esiukova E (2017) Microplastics in Baltic bottom sediments: quantification procedures and first results. Mar Pollut Bull 114(2):724–732. https://doi.org/10.1016/j.marpolbul.2016.10.060
72. Du J, Xu S, Zhou Q, Li H, Fu L, Tang J, Wang Y, Peng X, Xu Y, Du X (2020) A review of microplastics in the aquatic environmental: distribution, transport, ecotoxicology, and toxicological mechanisms. Environ Sci Pollut Res 27:11494–11505. https://doi.org/10.1007/s11356-020-08104-9

73. Yao L, Hui L, Yang Z, Chen X, Xiao A (2020) Freshwater microplastics pollution: detecting and visualizing emerging trends based on Citespace II. Chemosphere 245:125627. https://doi.org/10.1016/j.chemosphere.2019.125627
74. Zhao S, Zhu L, Wang T, Li D (2014) Suspended microplastics in the surface water of the Yangtze estuary system, China: first observations on occurrence, distribution. Mar Pollut Bull 86:562–568. https://doi.org/10.1016/j.marpolbul.2014.06.032
75. Han M, Niu X, Tang M, Zhang BT, Wang G, Yue W, Kong X, Zhu J (2020) Distribution of microplastics in surface water of the lower Yellow River near estuary. Sci Total Environ 707: 135601. https://doi.org/10.1016/j.scitotenv.2019.135601
76. Blair RM, Waldron S, Phoenix VR, Gauchotte-Lindsay C (2019) Microscopy and elemental analysis characterisation of microplastics in sediment of a freshwater urban river in Scotland, UK. Environ Sci Pollut Res 26:12491–12504. https://doi.org/10.1007/s11356-019-04678-1
77. Mani T, Hauk A, Walter U, Burkhardt-Holm P (2016) Microplastics profile along the Rhine River. Sci Rep 5:17988. https://doi.org/10.1038/srep17988
78. Faure F, Corbaz M, Baecher H, De Alencastro LF (2012) Pollution due to plastics and microplastics in Lake Geneva and in the Mediterranean Sea. ArchSci 65:157–164
79. Su L, Xue Y, Li L, Yang D, Kolandhasamy P, Li D, Shi H (2016) Microplastics in Taihu Lake, China. Environ Pollut 216:711–719. https://doi.org/10.1016/j.envpol.2016.06.036
80. Xiong X, Zhang K, Chen X, Shi H, Luo Z, Wu C (2018) Sources and distribution of microplastics in China's largest inland lake – Qinghai Lake. Environ Pollut 235:899–906. https://doi.org/10.1016/j.envpol.2017.12.081
81. Zhang K, Su J, Xiong X, Wu X, Wu C, Liu J (2016) Microplastic pollution of lakeshore sediments from remote lakes in Tibet plateau, China. Environ Pollut 219:450–455. https://doi.org/10.1016/j.envpol.2016.05.048
82. Lu H, Ziajahromi S, Neale PA, Leusch FDL (2021) A systematic review of freshwater microplastics in water and sediments: recommendations for harmonisation to enhance future study comparisons. Sci Total Environ 781:146693. https://doi.org/10.1016/j.scitotenv.2021.146693
83. Mbedzi R, Cuthbert RN, Wasserman RJ, Murungweni FM, Dalu T (2020) Spatiotemporal variation in microplastic contamination along a subtropical reservoir shoreline. Environ Sci Pollut Res 27(19):23880–23887. https://doi.org/10.1007/s11356-020-08640-4
84. Liu F, Olesen KB, Borregaard AR, Vollertsen J (2019) Microplastics in urban and highway stormwater retention ponds. Sci Total Environ 671:992–1000. https://doi.org/10.1016/j.scitotenv.2019.03.416
85. Li C, Busquets R, Campos LC (2020) Assessment of microplastics in freshwater systems: a review. Sci Total Environ 707:135578. https://doi.org/10.1016/j.scitotenv.2019.135578
86. Zhu F, Zhu C, Wang C, Cheng G (2019) Occurrence and ecological impacts of microplastics in soil systems: a review. Bull Environ Contam Toxicol 102:741–749. https://doi.org/10.1007/s00128-019-02623-z
87. Horton AA, Walton A, Spurgeon PJ, Lahive E, Svendsen C (2017) Microplastics in fresh water and terrestrial environments – evaluating the current understanding to identify the knowledge gaps and future research priorities. Sci Total Environ 586:127–141. https://doi.org/10.1016/j.scitotenv.2017.01.190
88. Fuller S, Gautam AA (2016) Procedure for measuring microplastics using pressurized fluid extraction. Environ Sci Technol 11:5774–5780. https://doi.org/10.1021/acs.est.6b00816
89. Zhou Y, Liu X, Wang J (2019) Characterization of microplastics and the association of heavy metals with microplastics in suburban soil of central China. Sci Total Environ 694:133798. https://doi.org/10.1016/j.scitotenv.2019.133798
90. Wang J, Li J, Liu S, Li H, Chen X, Peng C, Zhang P, Liu X (2021) Distinct microplastic distributions in soils of different land-use types: a case study of Chinese farmlands. Environ Pollut 269:116199. https://doi.org/10.1016/j.envpol.2020.116199
91. Cao L, Wu D, Liu P, Hu W, Xu L, Sun Y, Wu Q, Tian K, Huang B, Yoon SJ, Kwon B, Seong J (2021) Occurrence, distribution and affecting factors of microplastics in agricultural soils

along the lower reaches of Yangtze River, China. Sci Total Environ 794:148694. https://doi.org/10.1016/j.scitotenv.2021.148694
92. Liu M, Lu S, Song Y, Lei L, Hu J, Lv W, Zhou W, Cao C, Shi H, Yang X, He D (2018) Microplastic and mesoplastic pollution in farmland soils in suburbs of Shanghai, China. Environ Pollut 242(A):855–862. https://doi.org/10.1016/j.envpol.2018.07.051
93. Piehl S, Leibner A, Löder MGJ, Dris R, Bogner C, Laforsch C (2018) Identification and quantification of macro- and microplastics on an agricultural farmland. Sci Rep 8:17950. https://doi.org/10.1038/s41598-018-36172-y
94. Zhou Q, Zhang H, Fu C, Zhou Y, Dai Z, Li Y, Tu C, Luo Y (2018) The distribution and morphology of microplastics in coastal soils adjacent to the Bohai Sea and the Yellow Sea. Geoderma 322:201–208. https://doi.org/10.1016/j.geoderma.2018.02.015
95. Weber CJ, Opp C (2020) Spatial patterns of mesoplastics and coarse microplastics in floodplain soils as resulting from land use and fluvial processes. Environ Pollut 244:827–833. https://doi.org/10.1016/j.envpol.2020.115390
96. Scheurer M, Bigalke M (2018) Microplastics in Swiss floodplain soils. Environ Sci Technol 52(6):3591–3598. https://doi.org/10.1021/acs.est.7b06003
97. Beaurepaire M, Dris R, Gasperi J, Tassin B (2021) Microplastics in the atmospheric compartment: a comprehensive review on methods, results on their occurrence and determining factors. Curr Opin Food Sci 41:159–168. https://doi.org/10.1016/j.cofs.2021.04.010
98. Dris R, Gasperi J, Rocher V, Saad M, Renault N, Tassin B (2015) Microplastic contamination in an urban area: a case study in Greater Paris. Environ Chem. https://doi.org/10.1071/EN14167f
99. Dris R, Gasperi J, Saad M, Mirande C, Tassin B (2016) Synthetic fibers in atmospheric fallout: a source of microplastics in the environment? Mar Pollut Bull 104(1–2):290–293. https://doi.org/10.1016/j.marpolbul.2016.01.006
100. Cai L, Wang J, Peng J, Tan Z, Zhan Z, Tan X, Chen Q (2017) Characteristic of microplastics in the atmospheric fallout from Dongguan city, China: preliminary research and first evidence. Environ Sci Pollut Res 24:24928–24935. https://doi.org/10.1007/s11356-017-0116-x
101. Liu K, Wang X, Fang T, Xu P, Zhu L, Li D (2019) Source and potential risk assessment of suspended atmospheric microplastics in Shanghai. Sci Total Environ 675:462–471. https://doi.org/10.1016/j.scitotenv.2019.04.110
102. Abbasi S, Keshavarzi B, Moore F, Turner A, Kelly FJ, Dominguez AO, Jaafarzadeh N (2019) Distribution and potential health impacts of microplastics and microrubbers in air and street dusts from Asaluyeh County, Iran. Environ Pollut 244:153–164
103. Dris R, Gasperi J, Mirande C, Mandin C, Guerrouache M, Langlois V, Tassin B (2017) A first overview of textile fibers, including microplastics, in indoor and outdoor environments. Environ Pollut 221:453–458. https://doi.org/10.1016/j.envpol.2016.12.013
104. Liao Z, Ji X, Ma Y, Lv B, Huang W, Zhu X, Fang M, Wang Q, Wang X, Dahlgren R, Shang X (2021) Airborne microplastics in indoor and outdoor environments of a coastal city in Eastern China. J Hazard Mater 417. https://doi.org/10.1016/j.jhazmat.2021.126007
105. Vianello A, Jensen RL, Liu L, Vollertsen J (2019) Simulating human exposure to indoor airborne microplastics using a Breathing Thermal Manikin. Sci Rep 9:8670. https://doi.org/10.1038/s41598-019-45054-w
106. Klein M, Fischer EK (2019) Microplastic abundance in atmospheric deposition within the metropolitan area of Hamburg, Germany. Sci Total Environ 685:96–103. https://doi.org/10.1016/j.scitotenv.2019.05.405
107. Sridharan S, Kumar M, Singh L, Bolan NS, Saha M (2021) Microplastics as an emerging source of particulate air pollution: a critical review. J Hazard Mater 418:126245. https://doi.org/10.1016/j.jhazmat.2021.126245
108. Liu K, Wu T, Wang X, Song Z, Zong C, Wei N, Li D (2019) Consistent transport of terrestrial microplastics to the ocean through atmosphere. Environ Sci Technol 53(18):10612–10619. https://doi.org/10.1021/acs.est.9b03427

109. Wang Y, Huang J, Zhu F, Zhou S (2021) Airborne microplastics: a review on the occurrence, migration and risks to humans. Bull Environ Contam Toxicol. https://doi.org/10.1007/s00128-021-03180-0
110. Ryan PG, Moore CJ, Van Franeker JA, Moloney CL (2009) Monitoring the abundance of plastic debris in the marine environment. Philos Trans R Soc B Biol Sci 364:1999–2012. https://doi.org/10.1098/rstb.2008.0207
111. Bessa F, Ratcliffe N, Otero V, Sobral P, Marques JC, Waluda CM, Trathan PN, Xavier JC (2019) Microplastics in gentoo penguins from the Antarctic region. Sci Rep 9:14191. https://doi.org/10.1038/s41598-019-50621-2
112. Fang C, Zheng R, Zhang Y, Hong F, Mu J, Chen M, Song P, Lin L, Lin H, Le F, Bo J (2018) Microplastic contamination in benthic organisms from the Arctic and sub-Arctic regions. Chemosphere 209:298–306. https://doi.org/10.1016/j.chemosphere.2018.06.101
113. Iannilli V, Pasquali V, Setini A, Corami F (2019) First evidence of microplastics ingestion in benthic amphipods from Svalbard. Environ Res 179(Part A):108811. https://doi.org/10.1016/j.envres.2019.108811
114. Sequeira IF, Prata JC, da Costa JP, Duarte AC, Rocha-Santos T (2020) Worldwide contamination of fish with microplastics: a brief global overview. Mar Pollut Bull 160:111681. https://doi.org/10.1016/j.marpolbul.2020.111681
115. Ghosh GG, Akter SM, Islam RM, Habib A, Chakraborty TK, Zaman S, Enamul Kabir AHM, Shipin OV, Wahid MA (2021) Microplastics contamination in commercial marine fish from the Bay of Bengal. Reg Stud Mar Sci 44:101728. https://doi.org/10.1016/j.rsma.2021.101728
116. Hosseinpour A, Chamani A, Mirzaei R, Mohebbi-Nozar SL (2021) Occurrence, abundance and characteristics of microplastics in some commercial fish of northern coasts of the Persian Gulf. Mar Pollut Bull 171:112693. https://doi.org/10.1016/j.marpolbul.2021.112693
117. Sparks C, Immelman S (2020) Microplastics in offshore fish from the Agulhas Bank, South Africa. Mar Pollut Bull 156:111216. https://doi.org/10.1016/j.marpolbul.2020.111216
118. Rios-Fuster B, Alomar C, Compa M, Guijarro B, Deudero S (2019) Anthropogenic particles ingestion in fish species from two areas of the western Mediterranean Sea. Mar Pollut Bull 144:325–333. https://doi.org/10.1016/j.marpolbul.2019.04.064
119. Feng Z, Zhang T, Li Y, He X, Wang R, Xu J, Gao G (2019) The accumulation of microplastics in fish from an important fish farm and mariculture area, Haizhou Bay, China. Sci Total Environ 696:133948. https://doi.org/10.1016/j.scitotenv.2019.133948
120. Thiele CJ, Hudson MD, Russell AE, Saluveer M, Sidaoui-Haddad G (2021) Microplastics in fish and fishmeal: an emerging environmental challenge? Sci Rep 11:2045. https://doi.org/10.1038/s41598-021-81499-8
121. Meaza I, Toyoda JH, Wise JP (2021) Microplastics in sea turtles, marine mammals and humans: a one environmental health perspective. Front Environ Sci 8:575614. https://doi.org/10.3389/fenvs.2020.575614
122. Ugwu K, Herrera A, Gómez M (2021) Microplastics in marine biota: a review. Mar Pollut Bull 169:112540. https://doi.org/10.1016/j.marpolbul.2021.112540
123. Eerkes-Medrano D, Thompson RC, Aldridge DC (2015) Microplastics in freshwater systems: a review of the emerging threats, identification of knowledge gaps and prioritisation of research needs. Water Res 75:63–82
124. Teuten EL, Saquing JM, Knappe DRU, Barlaz MA, Jonsson S, Björn A et al (2009) Transport and release of chemicals from plastics to the environment and to wildlife. Philos Trans R Soc B 364:2027–2045. https://doi.org/10.1098/rstb.2008.0284
125. Carrasco-Navarro V, Muñiz-González AB, Sorvaria J, Martínez-Guitarte J (2021) Altered gene expression in Chironomus riparius (insecta) in response to tire rubber and polystyrene microplastics. Environ Pollut 285:117462. https://doi.org/10.1016/j.envpol.2021.117462
126. Détrée C, Gallardo-Escárate C (2018) Single and repetitive microplastics exposures induce immune system modulation and homeostasis alteration in the edible mussel Mytilus galloprovincialis. Fish Shellfish Immunol 2018(83):52–60. https://doi.org/10.1016/j.fsi.2018.09.018

127. Kim J, Yu Y, Cho J (2021) Toxic effects on bioaccumulation, hematological parameters, oxidative stress, immune responses and neurotoxicity in fish exposed to microplastics: a review. J Hazard Mater 413:125423. https://doi.org/10.1016/j.jhazmat.2021.125423
128. Wang Z, Fan L, Wang J, Xi J, Zhang C, Zhou C, Zhang L, Xu G, Zou J (2021) Insight into the immune and microbial response of the white-leg shrimp Litopenaeus vannamei to microplastics. Mar Environ Res 169:105377
129. Xia X, Sun M, Zhou M, Chang Z, Li L (2020) Polyvinyl chloride microplastics induce growth inhibition and oxidative stress in Cyprinus carpio var. larvae. Sci Total Environ 716:136479. https://doi.org/10.1016/j.scitotenv.2019.136479
130. Santos D, Luzio A, Matos C, Bellas J, Monteiro SM, Félixa L (2021) Microplastics alone or co-exposed with copper induce neurotoxicity and behavioral alterations on zebrafish larvae after a subchronic exposure. Aquat Toxicol 235:105814. https://doi.org/10.1016/j.aquatox.2021.105814
131. Jiang X, Chang Y, Zhang T, Qiao Y, Klobučar G, Li M (2020) Toxicological effects of polystyrene microplastics on earthworm (Eisenia fetida). Environ Pollut 240:639–646. https://doi.org/10.1016/j.envpol.2019.113896
132. Rodrigues JP, Duarte AC, Santos-Echeandía J, Rocha-Santos T (2019) Significance of interactions between microplastics and POPs in the marine environment: a critical overview. TrAC Trends Anal Chem 111:252–260. https://doi.org/10.1016/j.trac.2018.11.038
133. Kim SW, An Y (2020) Edible size of polyethylene microplastics and their effects on springtail behavior. Environ Pollut 266(1):115255. https://doi.org/10.1016/j.envpol.2020.115255
134. Andrady AL (2017) The plastic in microplastics: a review. Mar Pollut Bull 119(1):12–22. https://doi.org/10.1016/j.marpolbul.2017.01.082
135. Müller A, Becker R, Dorgerloh U, Simon F, Braun U (2018) The effect of polymer aging on the uptake of fuel aromatics and ethers by microplastics. Environ Pollut 240:639–646. https://doi.org/10.1016/j.envpol.2018.04.127
136. Pham DN, Cark L, Li M (2021) Microplastics as hubs enriching antibiotic-resistant bacteria and pathogens in municipal activated sludge. J Haz Mat Let 2:100014. https://doi.org/10.1016/j.hazl.2021.100014
137. Gkoutselis G, Rohrbach S, Harjes J, Obst M, Brachmann A, Horn MA, Rambold G (2021) Microplastics accumulate fungal pathogens in terrestrial ecosystems. Sci Rep 11:13214. https://doi.org/10.1038/s41598-021-92405-7
138. Parvin F, Jannat S, Tareq SM (2021) Abundance, characteristics and variation of microplastics in different freshwater fish species from Bangladesh. Sci Total Environ 784:147137. https://doi.org/10.1016/j.scitotenv.2021.147137
139. Lusher AL, O'Donnell C, Officer R, O'Connor I (2016) Microplastic interactions with North Atlantic mesopelagic fish. J Mar Sci 73:1214–1225. https://doi.org/10.1093/icesjms/fsv241
140. Amélineau F, Bonnet D, Heitz O, Mortreux V, Harding AMA, Karnovsky N, Walkusz W, Fort J, Grémillet D (2016) Microplastic pollution in the Greenland Sea: background levels and selective contamination of planktivorous diving seabirds. Environ Pollut 219:1131–1139. https://doi.org/10.1016/j.envpol.2016.09.017
141. Mercogliano R, Avio CG, Regoli F, Anastasio A, Colavita G, Santonicola S (2020) Occurrence of microplastics in commercial seafood under the perspective of the human food chain. A review. J Agric Food Chem 68(19):5296–5301. https://doi.org/10.1021/acs.jafc.0c01209
142. Nelms SA, Galloway TS, Godley BJ, Jarvis DS, Lindeque PH (2018) Investigating microplastic trophic transfer in marine top predators. Environ Pollut 238:999–1007. https://doi.org/10.1016/j.envpol.2018.02.016
143. Fernández B, Albentos M (2019) Insights into the uptake, elimination and accumulation of microplastics in mussel. Environ Pollut 249:321–329. https://doi.org/10.1016/j.envpol.2019.03.037

144. De Sales-Ribeiro C, Brito-Casillas Y, Fernandez A, Caballero MJ (2020) An end to the controversy over the microscopic detection and effects of pristine microplastics in fish organs. Sci Rep 10:12434. https://doi.org/10.1038/s41598-020-69062-3
145. Ragusa A, Svelato A, Santacroce C, Catalano P, Notarstefano V, Carnevali O, Papa F, Ciro M, Rongioletti A, Baiocco F, Draghi S, D'Amore E, Rinaldo D, Matta M, Giorgini E (2021) Plasticenta: first evidence of microplastics in human placenta. Environ Int 146:106274. https://doi.org/10.1016/j.envint.2020.106274
146. Amato-Lourenço LF, Carvalho-Oliveira R, Jumior GR, dos Santos Galvã L, Ando RA, Mauad T (2019) Presence of airborne microplastics in human lung tissue. Sci Total Environ 684:657–669. https://doi.org/10.1016/j.jhazmat.2021.126124
147. Hwang J, Choi D, Han S, Choi J, Hong J (2019) An assessment of the toxicity of polypro pylene microplastics in human derived cells. Sci Total Environ 684:657–669. https://doi.org/10.1016/j.scitotenv.2019.05.071

The manufacturer's authorised representative in the EU is Springer Nature Customer Service Centre GmbH, Europaplatz 3, 69115 Heidelberg, Germany. If you have any concerns regarding our products, please contact ProductSafety@springernature.com

Printed and bound by CPI Group (UK) Ltd, Croydon, CR0 4YY
30/09/2026
02225309-0007